Michael Gross
The Birds, the Bees and the Platypuses

Related Titles

Nicolaou, K. C., Montagnon, T.
Molecules that changed the World
2008
ISBN: 978-3-527-30983-2

Corey, E. J., Czakó, B., Kürti, L.
Molecules and Medicine
2007
ISBN: 978-0-470-22749-7

Emsley, J.
Better Looking, Better Living, Better Loving
*How Chemistry can Help You Achieve Life's
Goals*
2007
ISBN: 978-3-527-31863-6

Wagner, E. J.
The Science of Sherlock Holmes
*From Baskerville Hall to the Valley of Fear,
The Real Forensics Behind the Great
Detective's Greatest Cases*
2006
ISBN: 978-0-471-64879-6

Bell, H. P., Feuerstein, T., Güntner, C. E.,
Hölsken, S., Lohmann, J. K. (eds.)
**What's Cooking
in Chemistry?**
*How Leading Chemists Succeed in
the Kitchen*
2003
ISBN: 978-3-527-30723-4

Michael Gross
The Birds, the Bees and the Platypuses

Crazy, Sexy and Cool Stories from Science

WILEY-VCH Verlag GmbH & Co. KGaA

The Author

www.michaelgross.co.uk

Michael Gross
School of Crystallography
Birkbeck College
Malet Street
London WC1E 7 HX
United Kingdom

Library of Congress Card No.:
applied for

British Library Cataloguing-in-Publication Data
A catalogue record for this book is available from the British Library.

Bibliographic information published by the Deutsche Nationalbibliothek
The Deutsche Nationalbibliothek lists this publication in the Deutsche National-bibliografie; detailed bibliographic data are available in the Internet at http://dnb.d-nb.de.

© 2008 WILEY-VCH Verlag GmbH & Co. KGaA, Weinheim

Typesetting TypoDesign Hecker GmbH, Leimen
Printing and Binding Ebner & Spiegel GmbH, Ulm
Cover Design and Cover Illustration Himmelfarb, Eppelheim
www.himmelfarb.de

Printed in the Federal Republic of Germany

Printed on acid-free paper

ISBN: 978-3-527-32287-9

Contents

The Birds, the Bees and the Platypuses. Michael Gross
Copyright © 2008 WILEY-VCH Verlag GmbH & Co. KGaA, Weinheim
ISBN 978-3-527-32287-9

Preface

Science is fun! In seven years as a hobby reporter, and almost eight as a full-time freelance science writer, I have accumulated dozens of stories which I still remember fondly, because they were so much fun to write (and hopefully just as much fun to read). These are the stories that still tempt me to waste my time rereading them for the nth time if I stumble across them in my archives. These are the stories that I have used and reused over the years, cited as examples, or attached to my CV. These are the stories that – in my eyes, at least – demonstrate that science is a cultural activity just as rich and varied as literature and music, and just as rewarding.

What makes these stories stand out among the roughly 1000 others I have written over the years? I have identified three defining criteria, of which my favorite science stories may display one or more. Borrowing a title from TLC, I sorted them into a table with the headings crazy, sexy, and cool. *Crazy* stories include the weird, the unexpected, and the plain crazy stuff that scientists come across, and quite often discover to be actually useful. My favorite example of this kind are the wildly unorthodox antibodies found in camels and llamas, which have turned out extremely useful for biotechnology. There are also some stories of challenges so daunting that only crazy scientists would take them on. The genome sequence of our Neanderthal cousins springs to mind. *Sexy* stories are sometimes about sex (from attraction through to reproduction), but sometimes about other obsessions and characteristics of our race. Some of them just tell us what makes us human. *Cool* stories are mostly about cool inventions, devices, gizmos, and gimmicks. Many of them were invented by scientists, but there are also a few that were invented by evolution.

For each of these stories, I have started from a manuscript I wrote for publication (in a magazine or newspaper), revised and/or expand-

The Birds, the Bees and the Platypuses. Michael Gross
Copyright © 2008 WILEY-VCH Verlag GmbH & Co. KGaA, Weinheim
ISBN 978-3-527-32287-9

ed it, and added an introductory paragraph explaining what makes this particular story special. Where appropriate, I also attached an epilogue summarizing further developments. Within each of the three main sections, stories are arranged more or less chronologically, so one also gets a feel of how science has progressed in the years I've covered. At the bottom of each piece, the year of its first publication is given in brackets.

Many of these stories appeared originally in *Chemistry World*, the magazine of the Royal Society of Chemistry, or in its predecessor, *Chemistry in Britain*. A few articles from *Nachrichten aus der Chemie* (magazine of the German Chemical Society, GDCh), and *Spektrum der Wissenschaft* (the German edition of *Scientific American*), however, were published in German only, so I've translated them for this book. A couple of old *Spektrum* pieces reached this book via reflection by the earlier books *Life on the Edge*, and *Travels to the Nanoworld*. So it's all a big hall of mirrors, like the Y chromosome (page 38).

Some of these stories have also appeared in *Bioforum Europe*, *Bio-IT World*, *Current Biology*, *The Guardian*, *New Scientist*, *Süddeutsche Zeitung* and *Chemie in unserer Zeit*. I am grateful to all the editors who have commissioned my work over the years. Some of them developed the ability to read my mind, which can speed up the process and make life easier for me. But even when they ask challenging or really silly questions they help me to share my excitement with the readers.

Fifteen years is an extremely long time in scientific research, as I realized when editing the stories from the 1990s, some of which already had a somewhat historic, pre-genomic feel to them. Some of the things that I found exciting back then (and still do) appear to have fallen from fashion, while others have blossomed spectacularly. Some of the researchers involved have now got a Nobel Prize to their name; others appear to have disappeared from the radar. Such is life, even in science.

Above all, however, I am hoping to convey the impression that science in the last decade and a half was never boring, and that with every new answer that researchers work out, a host of new, even more exciting questions are likely to pop up, providing an endless supply of crazy, sexy, and cool findings.

Oxford, March 2008   *Michael Gross*

About the Author

Michael Gross was born in Kirn, Germany, but considers himself a European citizen. He began his writing career on the school's magazine, covering arts and humanities from Asterix to Picasso. As his Bohemian dreams of writing books in a Parisian café did not fulfil immediately, he opted for studying the sciences, and eventually managed combine his writing addiction and scientific training in a career as a full-time science writer after all not in Paris, but in Oxford. He earns his living mainly with the publication of articles in magazines, but also does some editing, translating, and lecturing, and occasionally writes entire books. Though his scientific interests span from quantum computation through to psycholinguistics, his heart-felt sympathy is with the strange creatures that live in volcanoes, the deep cold sea and hot geysers.

Michael Gross has been writing about science full time for the last eight years and as a night time hobby for the previous seven. From his treasure troves, he now presents his favourite science stories from these 15 years. What are the attractions that make him revisit a topic or reread an article again and again? Often, it's the sheer craziness of wildly unexpected findings or grotesquely oversized challenges. In other stories, there is a sexy element or a an unexpected insight into the human condition. And sometimes, when reporting new and future technologies, the author just can't help thinking: "cooooooool!" So here are xx crazy, sexy and cool science stories for you to enjoy.

The Birds, the Bees and the Platypuses. Michael Gross
Copyright © 2008 WILEY-VCH Verlag GmbH & Co. KGaA, Weinheim
ISBN 978-3-527-32287-9

1
Crazy Creatures

"If, at first, the idea is not absurd, there is no hope for it."

Albert Einstein

I have a natural tendency to favor slightly eccentric stories from science over the ones where a relevant question has been investigated and answered in a straightforward, almost predictable way. The craziness that interests me can arise from the random walks that evolution takes across time, or it may be found in the mind of the scientists who take on challenges so daunting that no sane person would bother with them. Or it could be both or somewhere in between. There is a whole spectrum of scientific craziness and crazy science.

But then again, some of the areas covered here started out as a blip of craziness in the margins of modern science but have since evolved to become mainstream research fields, possibly even with commercial potential. You never know what might happen, that's part of what makes eccentric topics so rewarding.

The Birds, the Bees and the Platypuses. Michael Gross
Copyright © 2008 WILEY-VCH Verlag GmbH & Co. KGaA, Weinheim
ISBN 978-3-527-32287-9

Squeezy Little Bears

The crazy creatures at the extreme ends of life on Earth have fascinated me for many years. As both my PhD thesis and one of my books dealt with life under extreme conditions, I'm no longer that easily impressed by tales of life in boiling water, sizzling deserts, or permanent ice. However, the following story (which unfortunately came up too late for the original edition of *Life on the Edge*) beats them all. If anybody wants to send animals to Mars, I suggest they try the "little bears" or tardigrades. The following text is adapted from a postscript included in the paperback edition of *Life on the Edge*.

Tardigrades are microscopically small animals reminiscent of downsized bears, at most half a millimeter long. They live in water droplets suspended in moss and lichens and can be found on all continents. Now if you're such a tiny little bear exposed to the elements, you need some very special survival skills.

Tardigrades have at least two major emergency routines. If their habitat is flooded and there is a risk of oxygen shortage, they inflate to a balloon-like passive state that can float around on the water for days. If, however, the threat comes from a lack of water, they shrink to form the so-called tun state (because it looks like a barrel), which could be described as the animal equivalent of a spore. Researchers have managed to resuscitate tardigrades by rehydrating moss samples after up to 100 years of storage on museum shelves, which proves the quite remarkable long-term stability of this state.

It was this tun state that Kunihiro Seki and Masato Toyoshima (Kanagawa University, Japan) used in their studies of resistance against high pressures. As the presence of water would have convert-

The Birds, the Bees and the Platypuses. Michael Gross
Copyright © 2008 WILEY-VCH Verlag GmbH & Co. KGaA, Weinheim
ISBN 978-3-527-32287-9

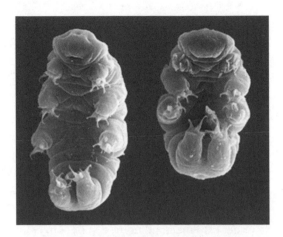

Figure 1 Electron micrograph of a tardigrade. Tardigrades or water bears are the most resistant animals known. (From: http://en.wikipedia.org/wiki/Image: Hypsibius-dujardini.jpg)

ed the animals back to the active state, the researchers suspended the tuns in a perfluorocarbon solvent before they applied pressures of up to 6000 atmospheres (more than five times the pressure found in the deepest trenches of the oceans). While active tardigrade populations in water are killed off by 2000 atmospheres (already an implausibly high threshold for an animal), the tun state allowed 95% of the individuals of one species and 80% of another to survive the maximum pressure of 6000 atmospheres.

This observation is unprecedented for any animal species. Only some bacterial spores and lichens could hope to compete with that. Still, tardigrade experts may have been only mildly surprised, as they knew already that the tuns can be revived after freezing in liquid helium – they are frost resistant down to 0.5 Kelvin. Detailed mechanistic explanations for these record-breaking achievements are not yet available. One thing that is known for sure is that the tuns contain high concentrations of the sugar trehalose, which is known to improve the stress resistance of baker's yeast.

The phenomenal shelf life of the tuns has aroused the interest of researchers in medical technology. Some are trying to copy the tardigrades' recipe to achieve similar long-term stability for human organs to be used in transplantation.

(2000)

Further Reading

M. Gross, *Life on the Edge*, Plenum, 1991.

What Happened Next

I am pleased to report that researchers actually followed up on my suggestion and sent tardigrades into space. The TARDIS (Tardigrades in Space) experiment was part of the FOTON M-3 mission, that launched on 14 September 2007 and returned safely on the 26th, after 189 orbits. At the time of writing, the tardigrade passengers were awaiting detailed analyses that will surely reveal how well they are suited to withstand space conditions.

http://tardigradesinspace.blogspot.com/

Can We Stomach the Bugs Bugging Our Stomachs?

The story of the bacteria that can give us ulcers has gone through many twists and turns over the years. Originally a heresy (everybody knew that ulcers are caused by acids!) the view that *Helicobacter* is bad for us soon turned into dogma (complete with recognition from Stockholm) which was challenged again by people suspecting that the bacteria may also have a beneficial effect. Here's just the basics, but I'll come back to the latest news on this towards the end of the crazy section.

When we prepare food, we often use extreme conditions such as high temperature or an acidic medium to kill off microorganisms. Industrial food preservation uses additional extremes including high pressure and gamma-ray sterilization. To a limited extent, our body can use similar methods. Thus, one of the functions of the acids in the stomach is to destroy bacteria taken in with the food.

However, scientists keep finding highly adapted microorganisms, known as extremophiles, thriving even under extreme conditions. For instance, hyperthermophiles can live at temperatures close to the boiling point of water, halobacteria can thrive in salt meat and *Deinococcus radiodurans* can survive gamma-ray sterilization. Similarly, there are other bacteria which can make themselves comfortable in the hostile environments of the mouth (like those which keep the dentists employed) and even in the stomach.

Much like the extremophile hunters in their field studies, the Australian pathologist J. Robin Warren found bacteria in a place where they should not be able to live according to textbook wisdom, namely in the human stomach. The spiral-shaped bacteria later classified as *Helicobacter pylori* had retreated into the mucus layer which covers the walls of the stomach. Only after a series of failures did Warren and his co-worker Barry J. Marshall manage to cultivate the new kind of bacteria. When they published their results in 1983, researchers around the world confirmed the occurrence of such bacteria in the stomach, especially in patients with chronic superficial gastritis, a condition involving a persistent inflammation of the stomach.

However, the presence of bacteria in a diseased tissue does not prove them guilty of causing the disease – they might just have profited from the body's weakness and invaded an organ already afflicted by disease. Marshall and a second volunteer did a self-test to find out whether the presence of the bacteria was the cause or a consequence of the disease. The two healthy men swallowed a dose of *Helicobacter pylori* and indeed both became ill with gastritis. Obviously, the infection with *Helicobacter* nearly always leads to a superficial gastritis, which, however, may often be overlooked and blamed on a heavy meal, for instance. If the infection persists and is not treated in time, it can lead to ulcers of the stomach or the duodenum in the long term.

This finding overturned a dogma that was almost as old as Western civilization, namely that ulcers are caused by excessive acid production by the stomach. In the first century AD, the Roman physician Celsus recommended low-acid food against ulcers. Since the 1970s, there have been drugs that reduce the acid production of the stomach without major side-effects, and indeed reduce ulcers. However, when the treatment is stopped, the ulcer always comes back. In contrast, treatment with bismuth prescriptions or antibiotics that eradicate the *Helicobacter* population can heal the gastritis permanently.

But how do the bacteria manage to settle in the human digestive tract without getting digested? The secret seems to lie in their mobility and in some chemical specialties of their metabolism. Mobility is crucial when the contents of the stomach get flushed down to the guts. With the help of their flagella, the spiral-shaped bacteria can swim fast enough to escape the fate of ending up in the loo. And the special "trick" of their metabolism is that they produce enormous amounts of the enzyme urease, which can degrade urea (a product of the digestion of proteins) to form ammonia and carbon dioxide. One possible explanation of the acid resistance of *Helicobacter* is that the bacteria may be able to use the ammonia produced by urease for the neutralization of the gastric acid in their immediate environment.

What looks like a crazy case of adaptation to extreme conditions is in fact immensely important for healthcare around the world. Scientists have estimated that a third of the world's population carries a latent infection with *Helicobacter* but, as with the tuberculosis germ, only a proportion of the infection cases lead to recognizable illness symptoms. About ten percent of all human beings develop ulcers at some stage of their life. Both for ulcers and for stomach cancer, a clear correlation with the number of individuals infected with *Helicobacter* can be found in comparative studies. Infections, and the diseases now believed to be their long-term consequences, are more common in developing countries than in industrialized ones, and they have both been declining slowly over the course of the twentieth century. A large-scale campaign to fight the germ could prove a very efficient measure against ulcers and cancers of the stomach and the duodenum.

(1996)

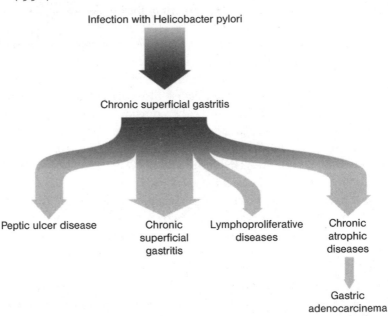

Infection with Helicobacter pylori

Chronic superficial gastritis

Peptic ulcer disease   Chronic superficial gastritis   Lymphoproliferative diseases   Chronic atrophic diseases

Gastric adenocarcinema

Figure 2 Consequences of an infection with Helicobacter. The relative widths of the arrows symbolize the different probabilities of the diseases concerned.

(Exzentriker des Lebens, 1997 © Spektrum Akademischer Verlag GmbH, Heidelberg, Spektrum Akademischer Verlag is a imprint from Springer SMB)

Further Reading

http://nobelprize.org/nobel_prizes/medicine/laureates/2005/index.html

What Happened Next

In 1997, Craig Venter's Institute for Genomic Research completed the genome sequence of *Helicobacter pylori*. Since then, plans have been mooted to eradicate the bacterium once and for all (the genome sequence would help to identify specific targets for a suitable drug), but some researchers have suggested that its ill effects on ulcers and some cancers may be counterbalanced by a more positive role in helping to avoid other cancers, so for the time being there is no systematic eradication program on the cards. In 2005, Barry Marshall and Robin Warren shared the Nobel prize for physiology or medicine "for their discovery of the bacterium *Helicobacter pylori* and its role in gastritis and peptic ulcer disease."

Oh, and in 2007 researchers found out something really crazy about *Helicobacter*, but I'll keep that for a later chapter in this section.

Protein, Edit Thyself!

Back in 1996, I wrote a story about the genome sequence of *Methanococcus jannaschii*, the first ever sequence of one of the weird bugs known as archaea, which are quite distinct from bacteria, even though this fundamental division was only discovered in the 1980s. I suggested using a circular graph that had appeared with the original paper, showing the entire genome at a glance, with different types of functional elements represented by different colors. As I'm not really responsible for the images that accompany my pieces, I didn't pay much attention to the details of this graph. Until, that is, the editor came back to me with the question: "What on Earth are inteins?" It turned out that in that image, 18 sequences were marked as inteins, and I didn't have a clue as to what they were. As it happened, they turned out to be really interesting, so I got to write a sequel to the genome story. By now, small genome sequences are no longer exciting, but inteins still are.

Imagine you bought a music cassette – yes, the ones with good old-fashioned magnetic tape in them – from a shop of magical music down at Diagon Alley. You take it home, open the box, and notice a loop of the tape hanging out of the cassette. The loop folds up in the shape of scissors and cuts itself loose from the rest of the tape. Not wanting to leave you with an unplayable cassette, though, the stray piece of tape now conjures up a small brush and glue, and sticks the loose ends of the remaining tape together, before it just wanders off. This may sound really crazy, but if you replace the magnetic tape with RNA or indeed with protein, it starts to make sense.

10 *The Birds, the Bees and the Platypuses.* Michael Gross
Copyright © 2008 WILEY-VCH Verlag GmbH & Co. KGaA, Weinheim
ISBN 978-3-527-32287-9

Inteins – self-splicing proteins – used to be considered as rather exotic, until 1996, that is. Then, in the genome sequence of *Methanococcus jannaschii*, the first genome from the domain of the archaea, 18 such sequences were discovered in 14 different genes, which more than doubled the number of known examples. In analogy to RNA introns, inteins can cut themselves free from the a longer polypeptide chain and link the remaining bits (called exteins in analogy with RNA exons). This process, however, was only discovered in 1990 and has not received as much attention as RNA splicing.

Although most of the known examples of inteins come from the domain of the notoriously idiosyncratic archaea (formerly archaebacteria), the prototype was discovered in a well-studied organism which has been of service to humankind for millenia: *Saccharomyces cerevisiae* (baker's yeast). Tom H. Stevens and his co-workers at the University of Oregon at Eugene observed that the TFP1 gene of yeast obviously codes for two protein products. The smaller one is coded in the middle region of the gene and flanked by the separate halves of the bigger one. This finding on its own would not have been very remarkable. There are many examples of overlapping or nested genes. It was unusual, however, that instead of the expected two messenger RNAs (one for each protein product) only one was found, and its length corresponded to the sum of the lengths expected. Stevens' group therefore suspected that the genetic information is not, as in many other cases, edited on the mRNA level. Rather, the single mRNA seemed to get translated into a single fusion protein, which splits into the final two components after translation.

To test this hypothesis, the researchers generated mutations in the middle part of the mRNA, leading to a shift of the reading frame, i.e. to a wrong segmentation of the string of nucleotides ("letters") into three-letter words specifying the amino acids to be incorporated into the protein. This kind of mutation not only affects the word where it occurs – the whole text behind it will be distorted as well. If there had been a splicing on the mRNA level, the frameshift would have only affected the intron cut out of the mRNA, as the spliced exons should still have had the correct frame. Therefore, only the protein coded by the middle segment should be mutated, not the one coded by the outer parts. It was found, however, that both proteins were affected by the frameshift. (Of course, one has to take care that the mutations of the middle part do not affect the splicing reaction, as can be confirmed by the molecular weights of the two products.) However, the researchers did not succeed in isolating the uncleaved precursor protein. This led them to suspect that the splicing might be an autocatalytic process. In this case, the protein itself would make the splicing reaction occur so rapidly that it would be impossible to get hold of the original translation product.

This difficulty was only overcome when inteins were also dicovered in several hyperthermophilic archaebacteria, such as *Thermococcus litoralis* and various species of *Pyrococcus*. Francine B. Perler and her co-workers at New England Biolabs constructed an artificial self-splicing system around the intein of *Pyrococcus* DNA polymerase by putting the gene for a maltose binding protein (M) in front of it (as a so-called N-extein, as it is at the amino-terminal end of the sequence) and a paramyosin gene (P) behind it (as a C-extein, for carboxy-terminal). They introduced this fusion gene into the intestinal bacterium *Escherichia coli* (a widely used laboratory workhorse), whose protein synthesis apparatus duly made the fused polypeptide (MIP) at temperatures between 12 and 32 °C. At these low temperatures, the self-splicing reaction occurred only very slowly, as the intein involved came from an organism adapted to life near the boiling point of water. In fact, the whole process was slowed down to such an extent that the researchers were able to purify the unprocessed precursor protein. Incubating this polypeptide in aqueous solutions containing only small amounts of sodium chloride and phosphate buffer, and then warming it up slowly, they could observe the onset of the self-splicing at higher temperatures. This way, they could also isolate an intermediate (MIP*), which behaved rather paradoxically. It appeared to have a higher molecular weight than MIP, as it moved more slowly through electrophoretic gels, and it also seemed to possess two different versions of the amino terminus (the "beginning" of a protein chain). The riddle was solved by the finding that the intermediate obviously has a branched structure, whose bulkiness decreased the mobility in gels. What must have happened is that the exteins M and P formed a link even while I was still attached to P.

In addition to their self-splicing abilities, inteins share a further characteristic with certain introns – or, more specifically, with certain proteins derived from intron translation. Both act as endonucleases, which means that they can recognize certain DNA sequences and cut the DNA at a well-defined position within or near these sequences.

They are specialized on DNA segments characteristic of their "home" gene but lacking the intein or intron sequence. Thus, they cut the gene at the corresponding position and thereby trigger a "repair" mechanism, which may use an intein/intron-containing copy of the gene as a template and thus produces the intervening sequence and inserts it into the gene.

This process – traditionally known as "intron homing" – has thus far only been demonstrated for four of the known inteins. Sequence comparisons, however, allow the conclusion that all known inteins are at least related to endonucleases, even if some of them may have lost the homing function in their recent evolution. The sporadic distribution of intein sequences among species of all three domains of life also suggests that this endonuclease activity has facilitated the spreading of inteins through horizontal gene transfer (i.e. transfer between contemporary species).

Why have self-splicing inteins and introns developed the function of homing-endonucleases? Well, maybe this question is put the wrong way round. If you turn it round, the answer is almost self-evident. Why are homing endonucleases self-splicing, be it on an RNA or on a protein level? An enzyme which is able to cut a gene and put its own genetic material right in the middle of it is potentially lethal for any cell, as it will at some time destroy an essential gene. If, however, the damage can be repaired on the RNA or on the protein level by means of the inserted sequence catalyzing a splicing reaction re-establishing the original product, all is well.

Thus, the intein or intron activity is the condition which a certain kind of mobile genetic element has to fulfill in order to be tolerated by the cell. If they don't fulfill it, they are damaging their own host cell and have only one chance of surviving – to acquire the ability to infect other cells and thus become viruses. Thus it is most probably no coincidence that self-editing on the protein level is most commonly observed in viruses. The genetic material of the AIDS virus HIV, for instance, codes for a long polypeptide chain containing all the viral proteins linked together. The HIV protease activity contained in this polyprotein cleaves the molecule into the desired proteins.

Perhaps the biological "purpose" of the inteins (which are obviously useless for the cell and only survive because they can help the spreading of their gene without hurting the cell too much) is to be understood as an evolutionary precursor or a peaceful alternative to viruses. One should definitely keep an eye on them.

(1996)

Further Reading

F. B. Perler and E. Adam, *Curr. Opin. Biotechnol*, 2000, 11: 377.
C. J. Bult *et al.*, *Science*, 1996, **273**, 1066.

What Happened Next

Inteins are now commercially available as part of a protein expression kit allowing researchers to produce a desired recombinant protein and to purify it in a single step. The idea is that a so-called affinity tag is encoded at one end of the gene, ensuring that the tail of the synthesized protein will have a strong and specific binding affinity to a certain material such as chitin. Thus it can be easily purified by running the cell extract through a separation column containing that material, where the protein of interest will bind and everything else will run through. An intein sequence located between the tag and the protein as such ensures that the tag is cut off after purification without the need for an additional procedure.

Magic Bullets from the Desert

This story began quite inconspicuously with a conversation I had in my office, back in the days when I was still doing research at the University of Oxford. A former colleague had come back to do some experiments in the course of a collaborative project that I didn't know anything about. When I asked him what his experiments were about, he said he was studying the interaction between lysozyme and camel antibodies. "Lysozyme" was the boring part of that sentence, as everybody in this lab had some connection to this classic workhorse of protein and enzyme studies. But the other half was news to me, so I pricked up my ears and asked: "Camel? What's special about camels?" So he told me, and I must have retold this story more than a dozen times in various formats. It's still one of my favorites.

It all started with a mutiny in a university teaching laboratory, some time in the late 1980s. A bunch of biology students were told to do the immunology experiments that countless others had done before them, fishing antibodies from human blood serum, and separating them into different groups. They were not too keen, as the serum might contain HIV, and also because the results of the experiment were well known and already documented in their textbooks. Their tutors offered to sacrifice a few mice instead – not a very popular choice either. Eventually, a few liters of serum leftovers were discovered in the freezers of the research labs – they were from dromedaries. This exotic sample inspired the students sufficiently to give up the strike action and start working on the separation of the antibodies. They found the usual distribution of immunoglobulins that one expects to

The Birds, the Bees and the Platypuses. Michael Gross
Copyright © 2008 WILEY-VCH Verlag GmbH & Co. KGaA, Weinheim
ISBN 978-3-527-32287-9

see, but they also discovered a group of smaller antibodies that did not correspond to anything known to science.

This episode happened at the Free University of Brussels, and it might have ended in obscurity had not two researchers at this university, Raymond Hamers and Cecile Casterman, investigated more deeply. They believed that that the smaller antibodies were not just degraded copies of the real ones, but that they were of a special kind. They repeated the students' experiments with fresh samples from camels and llamas, and confirmed that all the animals in this group (the Camelidae) produce some amounts of antibodies which are very different from the standard ones in that they are lacking the pair of protein molecules known as the light chains. They consist only of the heavy chains, which is why they are now referred to as heavy chain or HC antibodies. (In normal antibodies, a pair of heavy chains is arranged in a symmetrical Y shape, with one light chain attached to each of the branches, see Fig. 3.)

Ordinary antibodies are horrible things to deal with: they are complicated molecular assemblies, very difficult to produce in bacteria, too bulky for many medical applications, and may trigger an unwanted immune response in a patient (yes, there are antibodies against an-

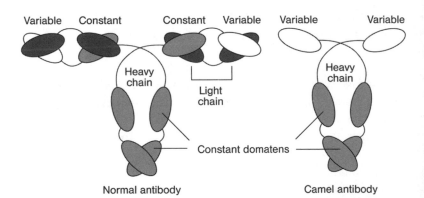

Figure 3 Camel antibodies. A "normal" antibody (left), of the kind typically found in vertebrates, represents a "Y" shape and consists of two heavy and two light chains. While the two heavy chains make up the foot of the Y, each of the arms consists of one heavy and one light chain. The antigen-binding site is equally composed of parts of the heavy and the light chain. The characteristic heavy chain antibodies found in camels and llamas (right) consist only of heavy chains. Thus, their binding sites consist of only one molecule and remain functional in the absence of the bulky remainder of the antibody.

tibodies!). Therefore, many research teams have tried to find something simpler, to construct miniature antibodies combining the binding specificity of a real antibody with some extra user-friendliness. Serge Muyldermans and Lode Wyns, also working at the Free University of Brussels, were pursuing research in this direction. They took up the trail of the camels, but it was quite a trek. In the beginning, it was far from clear that these molecules were functional antibodies with the high variability and specificity of the real thing. What they needed to do was to find a camel, immunize it with a specific antigen, wait a year, and see whether the camel had produced specific HC antibodies against that substance. A party of researchers traveled to Morocco, bought a camel, immunized it ... and had it stolen before they could get at the precious serum!

These practical problems were eventually overcome with a little royal help. His Highness, General Sheikh Maktoum bin Rashid Al Maktoum, then the ruler of Dubai in the United Arab Emirates, supported the research by providing camel serum from his renowned veterinary research centre. What the researchers found out about the camel antibodies was even more promising for medical and biotechnological applications than they could have hoped for.

It turned out that the HC antibodies, like normal ones, recognize a wide range of antigens, but they interact with them in different places. Thus, HC antibodies raised against small enzymes such as lysozyme or ribonuclease can penetrate the active site and provide a potent inhibitor for the enzyme, while conventional antibodies would bind somewhere more accessible.

This all boils down to the fact that in a conventional antibody, the antigen recognition is provided by two sites (at the upper ends of the "Y"), each composed of two different molecules, a heavy chain and a light chain, resulting in a rather bulky arrangement. In HC antibodies, each binding site is contained in a narrowly defined region of one molecule, the variable domain of the heavy chain. This is why it reaches the parts that other antibodies can't. This also means that it is a lot easier to further miniaturize this antibody. If you want to miniaturize a human antibody by cutting off all the parts that are not involved in binding, you get into enormous difficulties trying to keep the binding domains of the heavy and light chains together. With the camel version, you can just genetically isolate the DNA for the binding domain, get bacteria to make it, and you've got your miniature antibody, known to scientists as a single domain antibody.

Single domain antibodies are the ideal tool for a range of applications, from scientific research tools through to diagnostic kits to be used at home. One very promising field is imaging of living tissue, and especially cancer diagnosis. When trying to localize a tumor, you want a label that, apart from recognizing the particular molecules found on tumor cells, penetrates easily into the tumor. Once it has done that and bound to the target, you want to be able to wash out any remaining unbound material easily, so that it won't show up in the picture. Common antibodies fail these requirements, but preliminary tests suggest that single domain binders derived from camel HC antibodies could be used. It also appears that they will not normally provoke an immune response, as full-size (non-human) antibodies would. Furthermore, the small size of these molecules allows scientists to use them as building blocks for constructs which might contain two different binding sites, or even a binding site combined with an enzymatic or other activity. They could even be harnessed inside the cell, as so-called intrabodies.

A typical example of an antibody-based consumer product is the home pregnancy test kit which you dip into a urine sample, and then examine to see whether a blue stripe turns up. One version of this, designed to indicate the presence of a characteristic pregnancy hormone, contains two different kinds of antibodies against this hormone. One set is glued to the solid support in the window area where you want the blue stripe to appear. When hormone molecules float by, they will get bound by these antibodies. A second set of antibodies, recognizing a different part of the hormone molecule, is charged with blue-colored particles. When this second set comes across the hormone molecules firmly bound to the first set, it will bind to them and thus make the blue color accumulate in the window. As this kind of test requires two kinds of antibodies that bind the target in different ways such that they do not interfere with each other's binding, a combination of conventional antibodies on the solid surface and camel-type antibodies in the liquid would be ideal.

(2000)

Further Reading

S. Muyldermans *et al.*, *Trends Biochem. Sci.*, 2001, **26**, 230–235.

What Happened Next

Since 2002, a spin-out company called Ablynx has begun to develop a number of products based on the advantages offered by camelidae antibodies. As of 2007, Ablynx has over 90 staff and contracts with several pharma giants. The first drug based on camel antibodies has just entered a phase I clinical trial. Watch this space.

Further Reading

T. N. Baral *et al.*, *Nature Med.* 2006, **12**, 580.

Asparagine and Old Lace

As a teenager, I was an avid reader of popular mathematics, including the books of *Scientific American* columnist Martin Gardner. Thus, anything that can link this world to the one I later came to work in, namely proteins, is bound to tickle me. Protein scientists have a bad habit of misusing the word "topology" when they simply refer to structure. However, in this mathematics/protein science crossover, there is some real topology involved, as protein chains get tied into knots.

Supramolecular chemists have come up with clever ways of producing molecular assemblies such as knots and interlocking rings (catenanes). This appeared to be a playful, sometimes even useful, new invention without much of a model in nature, apart from DNA which can be tied up in topologically tricky states owing to its double helix packing and its habit of forming rings. Proteins, the structurally most diverse biopolymers, were assumed to stay straight-laced as a matter of principle. In 2000, however, natural proteins were shown to form both knots and catenanes.

John Johnson and his coworkers at the Scripps Research Institute reported in *Science* the high-resolution structure of the protein "head" of a virus called HK97. While most of the protein structures determined nowadays yield only variations on well-known themes, this one held two major surprises. The protein molecule in itself adopts a three-dimensional structure of a type ("fold") not observed before. But what happens between the 420 protein molecules that make up the shell was even more surprising. Once they have assembled in the ball-shaped structure, they are not content to stick together by weak, non-covalent bonding, as protein subunits of complex assembly systems

The Birds, the Bees and the Platypuses. Michael Gross
Copyright © 2008 WILEY-VCH Verlag GmbH & Co. KGaA, Weinheim
ISBN 978-3-527-32287-9

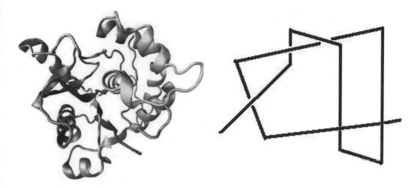

Figure 4 Protein in a tangle. Researchers used advanced computation methods to "pull" both ends of a protein chain and find out whether it is knotted or not. In this case, a knot was revealed.

normally do. Instead they form a new amide bond from the primary amine of a lysine residue of one unit and the amide of the asparagine of its neighbor.

Once this has happened 420 times, each of the protein molecules finds itself not only covalently integrated into a ring with four or five others. It is also irremovably chained to its neighboring rings, and thus to the entire assembly. Thus, 60 hexameric and 12 pentameric rings form a three-dimensional meshwork that can only be opened by breaking chemical bonds. This explains the remarkable stability of the shell, economically achieved with less protein than would have otherwise been necessary.

Less than a month earlier, William Taylor of the National Institute for Medical Research at Mill Hill, London, demonstrated the existence of strongly knotted protein chains by combing through known structures with a new algorithm he had developed, essentially holding the ends of the protein chain tight and then shrinking the chain. Most chains will end up as a straight line between the locations of the termini, but some turned out to be knotted. The most complex knot observed was a figure-of-eight knot in the plant protein acetohydroxy acid isomeroreductase – now say this very quickly and observe how your tongue gets knotted up to a figure-of-eight!

Further Reading

W. R. Taylor and K. Lin, *Nature*, 2003, **421**, 25.

What Happened Next

In 2006, researchers at MIT reported that they found the most complex knot ever seen in a protein. After pulling the strings of more than 30,000 protein structures, they found several knots with three or four crossings, but just one had the record-beating five crossings (Fig. 4). The tangled protein is the enzyme ubiquitin hydrolase, which plays an important part in labeling proteins destined to be degraded by the cell's protein recycling machine, the proteasome. Specifically, the hydrolase removes the ubiquitin label and thus rescues proteins from certain death. The MIT researchers have speculated that the complex knot may be a special protective gear that stops the enzyme, which is so closely associated with the death marker ubiquitin, from ending up in the recycling bin.

Better Reasons to Kiss a Frog

Back in the days when I was a wide-eyed PhD student, we used to have a literature seminar including long-ish presentations on topics that could very well be quite a way off the usual preoccupations of our laboratory. I remember one of the speakers announcing that he was going to talk about peptides from frog skin, and I remember very clearly thinking "What the **** is that about?" By the end of the seminar, however, he had convinced me that these peptides were really worth knowing about. So much so that when, a few years later, similar peptides were discovered in human skin, I came prepared.

For many centuries, the savants of native populations in both Africa and South America have prescribed remedies made from frog skin against many ailments, and we can now understand why. The skins of frogs and toads contain a range of pharmacologically active substances, including antibiotics. In the late 1980s, the group of Michael Zasloff discovered the first animal-produced broad spectrum antibiotic in the skin of the clawed toad *Xenopus laevis*. It turned out to be a short peptide and was called magainin (after the Hebrew word for shield). Mammals including mice and humans, however, initially seemed to lack this shield.

At present, there are two groups of antibacterial peptides known to science that also occur in mammalian skin: the beta defensins, and the cathelicidines. Certain cells of the skin and of the immune system produce these peptides by cutting them out of larger precursor proteins. The observation that they kill bacteria in the test tube, however, does not strictly count as proof of their biological function as an an-

The Birds, the Bees and the Platypuses. Michael Gross
Copyright © 2008 WILEY-VCH Verlag GmbH & Co. KGaA, Weinheim
ISBN 978-3-527-32287-9

tibiotic in the skin. This latter function was first demonstrated in 2001.

Richard Gallo and his group at the University of California at San Diego turned their attention to a mouse protein named CRAMP, which was known to be closely related to the human protein LL-37, both belonging to the family of the cathecilidins. To study the role of this protein in the interaction between the mouse and potentially harmful bacteria, the researchers undertook elegantly symmetric studies involving genetic alteration of the mouse in one experiment, and of the bacteria in another.

First they bred mice totally devoid of this gene (knockout mice). When grown in an aseptic environment, these mice are normal and healthy. After an artificial infection with streptococci, however, the CRAMP-deficient mice were more seriously affected than the normal animals of the wild type control group. In the complementary experiment, the researchers produced bacteria that are resistant to CRAMP. They duly found the altered staphylococci to be more pathogenic to normal mice than the normal, CRAMP-sensitive bacteria.

Other researchers discovered a novel peptide antibiotic, neither a defensin nor a cathelicidin, also produced in the skin but ending up in the sweat. The group of Birgit Schittek at the University of Tübingen, Germany, started by investigating the properties of a gene from the human genome, whose function they did not know. They screened dozens of human tissues without finding any activity of this gene, and eventually found that it is expressed exclusively in one particular type of sweat gland. It codes for a small protein of just 110 amino acids which the researchers called dermicidin. In the sweat, however, they only found a fragment from the carboxy terminus of this protein, a 47-residue peptide they called DCD-1.

They synthesized an analogue of DCD-1 and investigated its antibacterial effects in comparison to an N-terminal fragment of dermicidin and an unrelated peptide. Only the DCD-1 analogue proved to be lethal, being effective not only against bacteria such as *Escherichia coli* or *Enterococcus faecalis* but also against the ubiquitous infectious yeast *Candida albicans*, suggesting a remarkably broad activity spectrum. The analogue displayed this activity both in standard assay buffers and in a liquid designed to mimic the chemical composition of human sweat.

Details of the molecular structure and biological role of the sweat peptide remain to be explored. It may be involved in controlling the microbial flora on our skin, or it may be limited to protecting the sweat glands from invasion by skin microbes. Knowledge of its gene sequence will enable researchers to go fishing for related substances which may have escaped discovery so far. Considering the important role of the skin in fighting off microbial invasion from the outside world – not to mention the effort human inventiveness has put into making it look good and healthy – we really should have a better understanding of what is going on at this frontier.

Further Reading

C.L. Bevins and M. Zasloff, *Annu. Rev. Biochem.*, 1990, **59**, 395–414.
V. Nizet *et al.*, *Nature*, 2001, **414**, 457.
B. Schittek *et al.*, *Nature Immunol.*, 2001, **2**, 1133–1137.

What Happened Next

In 2006, researchers in Israel created a new kind of miniature antimicrobial peptide combining features of two different kinds of antimicrobials in a minimalist design. The new agents contain only four amino acid residues, combined with a fatty acid.

Yechiel Shai and his coworkers at the Weizmann Institute at Rehovot, Israel, created their "ultrashort lipopeptides" as a synthetic alternative to two groups of antimicrobials found in nature, namely the antimicrobial peptides (AMPs), which are typically made up of 12–50 amino acids and carry a net positive charge, and lipopeptides, combining a lipophilic chain with a short anionic peptide of six or seven amino acid residues.

Exploring the minimum requirements for such peptides still to exhibit antimicrobial activity, the researchers were surprised to find that even peptides with only four amino acids, when combined with a fatty acid with a chain length of 12 to 16 carbon atoms, showed activity against bacteria and/or fungi. For example, a peptide of the sequence lysine-glycine-glycine-D-lysine, with a hexadecanoic acid (16 carbon atoms) attached to the amino terminus turned out to be the most potent compound. As the link between the lipophilic part and the peptide is just another peptide bond, the whole compound can be prepared on a commercial peptide synthesizer using established procedures.

Despite the small size of these molecules, the researchers established that they disrupt the cell membrane in similar ways to the larger lipopeptides and AMPs. For example, the uptake of a green fluorescent dye through the normally impermeable membrane is counted as evidence for membrane damage caused by the lipopeptides.

Interestingly, both the choice of amino acids and the length of the lipid chain can be used to fine-tune the specificity of the agents.

Further Reading

A. Makovitzki, A. Avrahami, and Y. Shai, *Proc. Natl. Acad. Sci. USA*, 2006, DOI 10.1073/pnas.0606129103.

Health Warning: Your Body May Be Unstable

The textbooks say that evolution twiddled the stability of our most abundant structural protein such that its melting point is just a few degrees above body temperature. A closer look at this seemingly simple issue has revealed that collagen actually melts just below physiological temperature. Should we be worried?

A cherished dogma from the early days of molecular biology stated that the biologically active state of a protein molecule is identical with a unique conformation intricately folded up in space, the so-called native state. Pull this apart, and you get something disordered, ill-defined, of no use whatsoever, the random coil. Or so we believed. In the late 1990s, however, researchers learned that the unfolded state isn't that random and horrible after all, and that it can even have a biological function. A growing list of proteins is known to defy dogmatic thinking by existing in a clearly unfolded state under physiological conditions. In many of these cases, this flexible state can adopt a second, more-folded conformation after encountering a ligand or receptor molecule. Protein–ligand interactions can be much tighter if the protein folds "around" the ligand than if they meet with a predefined key/lock complementarity.

To the list of around 100 proteins that are most likely to be unfolded under conditions corresponding to their physiological environment, researchers may have to add one that will make an unlikely candidate, as it is an important part of the very fabric of the human body, which we would, of course, like to think of as ordered, functional and stable. This protein, collagen, making up your tendons, skin, fingernails, and many of the less-visible structural connections in the body, accounts for more than one-third of the protein found in humans and all other vertebrates. And yet, a very diligently designed calorimetric study by the group of S. Leikin at the NIH (Bethesda, Maryland) showed that the thermodynamically stable state of the collagen variant from human lungs at physiological temperature (37 °C) is the unfolded state.

Essentially, the researchers realized that previous studies of the heat-induced unfolding of collagen had been conducted with too rapid heating. Before the slow unfolding equilibrium would have adjusted to one temperature step, researchers would already have applied the next step, thus obtaining an apparent transition temperature several degrees above the true value. Leikin's group found that the transition is so extremely slow that it doesn't even equilibrate in time when the slowest possible heating rate (0.004 °C/min) is used. They used extrapolation to find the "true" melting point of human lung collagen and determined it to be 36 °C. So are we all on the edge of a body meltdown? Is our lifetime borrowed from a kinetic anomaly?

Although some of our collagen does indeed get lost through unfolding, we shouldn't worry too much about this finding. Similarly to the above-mentioned binding proteins which are "natively unfolded" in solution, all the better to bind their ligands when they find them, collagen does get a tighter grip when it finds molecules of its own kind and forms the collagen fibrils which are the very fabric of many of our tissues. Thus, the instability only affects the collagen molecules for a limited time between secretion from the cell and formation of the mature fibers.

In the cell, a collagen-specific molecular chaperone, the heat shock protein Hsp47, takes care of the procollagen monomers. Once they have been secreted, however, they are no longer chaperoned and their stability seems to benefit very little from the presence of the short prosequences which distinguish the procollagen (precursor protein) from the mature product. Leikin and his team conclude that at this stage partial unfolding must occur, which is consistent with previously postulated "micro-unfolding" as part of the aggregation into collagen fibers. After removal of the propeptides, the flexibility of the micro-unfolded chain facilitates the formation of fibers, and also the removal of those molecules that are not incorporated and have to be degraded.

Most reassuringly, the thermodynamical situation improves dramatically once the protein molecules are safely ensconced in collagen fibrils. Although a certain amount of micro-unfolding may still occur and contribute to the elasticity of the structure, the limited space available to each molecule implies that the entropic gain of the unfolding process is reduced, hence the melting point increased. All in all, evolution has adjusted the balance of flexibility and stability just right, even though finding this balance isn't quite as simple as we used to think.

Further Reading

K.W. Plaxco and M. Gross, *Nature*, 1997, **386**, 657.
V.N. Uversky *et al.*, *Prot. Struct. Funct. Genet.*, 2000, **41**, 415.
K.W. Plaxco and M. Gross, *Nature Struct. Biol.*, 2001, **8**, 659.
E. Leikina *et al.*, *Proc. Natl. Acad. Sci. USA*, 2002, **99**, 1314.

What Happened Next

Rumors of an imminent structural collapse of all humans on Earth have been greatly exaggerated and thus failed to materialize. However, the concept of proteins that are intrinsically unfolded and only adopt a folded structure upon binding or under specific environmental conditions has proven very useful for the understanding of a wide range of biological phenomena, including some of medical relevance.

All Together Now

I have always enjoyed challenging the traditional view that humans are the "highest" life form on Earth and everything else is more or less primitive. For instance, bacteria aren't quite as dumb as they may appear. Large numbers of them can coordinate their activities in time and space, relying on a form of chemical communication known as quorum sensing. Social behavior in bacteria? What a crazy idea ...

An old but persistent myth has it that our own species *Homo sapiens sapiens* represents the most sophisticated life form on Earth, while bacteria are the most primitive one. Douglas Adams has famously overturned the first part of this myth, and microbiologists are now beginning to realize that bacteria are a lot more sophisticated than we mere mortals used to think. True, the few species that can be easily cultivated in the laboratory are often well described by the assumption of a single cell programmed to divide after a certain time as long as there is enough food around. But if you observe bacteria out in the wild, they are a lot more complex and less predictable.

Microbiologists now estimate that the majority of the bacterial biomass is in fact not found as free-living single cells, but rather involved in some kind of higher organization, including symbiosis with other organisms (e.g. in lichens or animal guts) and biofilms. Cyanobacteria can become part of lichens, which look deceptively like higher plants, and also in layer structures that display strict structural organization on large scales. Colonies of luminescent bacteria can send out precisely coordinated flashes of light. All these "social" activities require each individual bacterium to know of the presence of the others and to communicate with them. For this purpose, they have a chemical signaling system known as quorum sensing.

This was originally discovered in luminescent bacteria, which only light up when there are many of their friends around. In the 1970s, researchers showed that the bacteria secrete a molecular messenger, called the autoinducer, into the medium, and only produce light when they sense a threshold concentration of this molecule. For many years, biologists believed this communication to be specific to bioluminescence. It was only in the 1990s that quorum sensing turned out to be a much more general phenomenon, involved in disparate processes including synthesis of antibiotics in *Erwinia carotovora*, and the production of virulence factors in pathogenic bacteria.

The molecular mechanisms of quorum sensing have long remained mysterious. In 2002, two crystal structures of proteins involved in the process allowed researchers to put together at least some of the fundamental pieces of the mechanism. The group of Frederic Hughson at Princeton University (New Jersey) identified a hitherto elusive autoinducer known as AI-2 by solving the crystal structure of its receptor, which turned out to contain the AI-2 molecule. While many of the bacterial pheromones known so far are specific to one species, AI-2 appears to be widely distributed and might even serve as a communication device between different species.

The receptor in question was LuxP, a protein involved in the coordinated bioluminescence of the marine bacterium *Vibrio harveyi* (a harmless distant relative of the cholera germ, named after the pioneer of bioluminescence research, E. Newton Harvey). While the protein structure as such was similar to those of other binding proteins located in the periplasm (the space between cell membrane and cell wall), it was the unusual chemical structure of the autoinducer trapped inside, representing the first example of a biomolecule containing boron, which secured its place on the pages of *Nature*.

A few months later, the group of Andrzej Joachimiak at the Argonne National Laboratory (Argonne, Illinois) presented another crystal structure of a key protein involved in quorum sensing. Their target is the protein TraR from the plant pathogen *Agrobacterium tumefaciens*. This protein is related to another quorum sensor from the bioluminescence system and constitutes a direct link between pheromone recognition and the resulting change in gene expression, as it acts both as a signal receptor and as a transcription enhancer. The

Argonne group managed to catch it *in flagrante*, with two molecules of the autoinducer and a piece of the target DNA bound to the protein dimer.

One of the most intriguing aspects of the resulting structure is that the pheromone appears to be completely encapsulated within the protein fold. In accordance with earlier biochemical work indicating that the protein acquires resistance against protease digestion when binding the small molecule, this finding suggests that the sensor "folds around" its messenger molecule. In other words, it starts out from some more-loosely folded, probably monomeric, conformation, and only folds into the DNA-binding dimer when it has secured its two molecules of the autoinducer. Thus, binding of the signaling molecule is an essentially irreversible switch from an inactive TraR to the active conformation.

As *A. tumefaciens* makes its living by invading plants and setting up colonies in structures which look like tumors (hence the name), it is rather important for the individual bacteria to know whether they are part of a successful invasion troop, or whether they are out on their own. The traR gene is switched on as soon as the bacterium senses certain plant-specific chemicals, and the individualist turns into a part of a coordinated army from then on. Deeper understanding of bacterial communication, gathered from the present structural work and future research, should hopefully enable us also to fight bacterial invasions of our own bodies more efficiently – seeing that we are supposedly smarter than they are.

(2002)

Further Reading

The quorum sensing site: http://www.nottingham.ac.uk/quorum/
X. Chen et al., Nature, 2002, **415**, 545.
R. Zhang et al., Nature, 2002, **417**, 971.

What Happened Next

No bacterial communications have reached my sensors in recent years, but I am sure that a lot of interesting work continues to be performed in this field – try visiting the quorum sensing site to get an impression of what's going on.

So Where is Most of the Universe?

Nothing quite winds me up like the sound of people pretending that they understand the Universe. Because if they were honest, they would have to admit that they don't have the foggiest idea about some 95% of the mass/energy content of the Universe. Cosmologists invented the labels "dark matter" and "dark energy" in order to refer to those parts they knew nothing about, but that's as far as they got. In 2003, I attended a Royal Society discussion meeting on this topic and came away traumatized by the sheer enormity of our ignorance.

When theoreticians get stuck, they sometimes resort to the emergency exit of postulating a new particle or substance which cannot be observed by current methods. Sometimes it turns out to be non-existent (like, most famously, phlogiston), but sometimes it can be tracked down decades later, like the neutrino, which Wolfgang Pauli introduced very reluctantly in 1930. Similarly, the dark matter problem has generated a variety of hypothetical particles and on top of that adopted a few more, including Pauli's neutrino, to fix that giant hole in the Universe. Here is a short overview of past and present candidate explanations:

- There is no dark matter, and the observations are explained by Modified Newtonian Dynamics (MOND). This school of thinking (dismissed by most researchers in the field) claims that Newton's law of gravitation might be different over very long distances. While the MOND theory can explain a few observations at the cost of replacing Newton's law with a less elegant equation, it does not consistently explain the entire Universe.

The Birds, the Bees and the Platypuses. Michael Gross
Copyright © 2008 WILEY-VCH Verlag GmbH & Co. KGaA, Weinheim
ISBN 978-3-527-32287-9

- "Baryonic dark matter." In the early days of dark matter research, it was considered possible that the missing mass might consist of ordinary matter after all, only hidden in unexpected places, such as dwarf stars, planets, or gas clouds. Now, however, a wealth of results including isotope abundance, gravitational lensing, and the microwave background has proven unequivocally that the dark matter must be in non-baryonic particles. Most crucially, baryonic matter would not be able to account for the large-scale structures (galaxies upwards) we see in the Universe.
- Neutrinos. After many years of uncertainty, researchers have recently shown that neutrinos have a small mass. Thus they could contribute to the dark matter, especially if there are other kinds of neutrinos apart from the three detectable ones. In this context, they are known as "hot dark matter." So far, however, dark matter that is predominantly hot cannot explain the structure of the Universe.
- Warm dark matter, e.g. gravitinos, which arise from the theory of supersymmetry as partners of the graviton. Currently thought to be among the less-promising candidates.
- WIMPs are "Weakly Interacting Massive Particles," representing an entire class of possible constituents of cold dark matter, including neutralinos (postulated from supersymmetry theory). Several facilities have been built with the specific purpose of finding them.
- Axions. First postulated to plug a gap in the Standard Model of particle physics, axions entered the dark matter field when simulations predicted that their overall abundance in the Universe should be very high if their mass is small enough. There is also the idea of large-scale axion clusters, which may be distributed over light years of space.
- Cold thermal relics. Without the need to specify which kind of particle it refers to, this theory assumes that the present non-interacting cold matter was once in thermal equilibrium with the rest of the world. At one point, it decoupled and went its separate ways. This theory has proven useful for predictions even though it does not tell us what particles the dark matter is made of.

There are many more candidate ideas, and they come in all shapes and sizes. As Rocky Kolb pointed out ironically, the mass has been "pinned down to within 65 orders of magnitude," and the interaction characteristics range from absolutely non-interacting (apart from gravitation) through to strongly interacting.

How can researchers make sense of all this and pick the right explanation from such a large range of possibilities? There are some who hope that fundamental physics will one day come up with a Theory of Everything, which will, by definition, explain the entire Universe including dark matter and dark energy. More likely, the solution will come from a lot more observations and experiments, and a lot of hard work.

Dark Energy

Astronomical observations allow us to observe the past, because the light from distant stars needs millions of years before it arrives in our telescopes. Moreover, the distribution of matter and radiation in the Universe allows researchers to retrace its evolution and to describe in reasonable detail what happened between the Big Bang and now. There is, however, no window into the future of the Universe, and at the moment "what happens next" appears more uncertain than ever. The problem is that the future of the Universe depends very sensitively on the nature of the dark energy.

The previous paradigm that the Universe is dominated by gravitating matter (no matter whether it's dark or luminous) and will after a period of expansion collapse back in a Big Crunch event was shattered by relatively simple astronomical observations made on supernovae of the type Ia. Essentially, these supernovae can serve as a beacon because they are all extremely similar in their intensity and spectral characteristics. Thus, from the apparent magnitude with which a distant supernova is observed, astronomers can deduce its distance in time and space. From the wavelength shift towards the red that the light has suffered in transit they can calculate how much the Universe has expanded during that time.

This is a surprisingly simple way of monitoring the expansion history of the Universe. Saul Perlmutter and his colleagues used this approach in the mid-1990s, expecting to find out how rapidly the expansion of the Universe is slowing down. It took only a few dozen supernovae to convince them that their fundamental assumption was wrong and to prove that the expansion rate of the Universe is currently accelerating.

Combined with the current knowledge of ordinary and dark matter, and the cosmic microwave background, this insight leads directly to the conclusion that two-thirds of the mass/energy content of the Universe must be tearing things apart instead of pulling them together, as gravitating matter would. It does not tell us yet what the nature of this dark energy is, and whether it is invariable (cosmological constant) or changes over time (e.g. quintessence). If it is a constant vacuum energy, there is the dilemma that it would have to be many orders of magnitude different from what particle physicists predict it to be, and the intriguing coincidence that a parameter set at the birth of the Universe to a value insignificantly small at that time should begin dominating cosmology exactly at the time when we turn up to see it happening. On the other hand, allowing the dark energy to change over time throws up even more questions and new variables that we know nothing about. Can supernova observations be improved to reveal more about the dark energy mystery?

At present, data from supernovae in the significant distance range are building up quickly. Soon there will be hundreds of them. At that point, the information to be gained is no longer limited by statistical error, but by the risk of systematic error. There are three kinds of significant errors that need to be controlled: Firstly, comparing supernovae from different epochs in the history of the Universe carries the risk that the average properties of these events might have shifted over time. For instance, the age of the star at explosion, or the elemental composition might be different in an average modern supernova, if compared to average early supernovae. Luckily, the spectroscopic analysis of the light obtained from the supernovae allows the observer to compile a complete picture of their physical properties. Making sure that they have large numbers of well-characterized supernovae and then only compare like with like, astronomers can rule out any adverse effect of time differences.

Secondly, interstellar dust in the light path might distort the observation of distant events. To rule this out, observers must perform control experiments, for example checking for scattering of the light from nearby objects in other parts of the spectrum, particularly for X-ray sources. Lastly, very distant events might be amplified by gravitational lensing, so a careful observation of the environment is required to account for this possibility.

In the short-term, researchers are addressing these issues with a large-scale project called the Nearby Supernova Factory, which will record complete high-precision datasets of hundreds of supernovae every year. To look further back in time and be able to monitor the time when the Universe switched from slowing down to accelerating (i.e. when dark energy gained the upper hand over gravitating matter), researchers are planning to install a space-based telescope and spectrometer dedicated to this task and called SNAP (SuperNova Acceleration Probe). This probe, used in collaboration by many research groups mainly based in the US and France, will monitor thousands of supernovae in unprecedented detail.

Selecting suitable subsets from this large number of supernovae, researchers will be able to reduce the error bars significantly and thereby narrow down the range of possibilities currently open to the nature of the dark energy and the fate of our Universe. In combination with continuing progress in other experimental observations, including the probing of the microwave background, mass density, and gravitational lensing, they might even approach definitive answers to cosmology's most puzzling mysteries.

(2003)

Further Reading

K.W. Plaxco and M. Gross, *Astrobiology: A brief introduction*, JHU Press, 2006.

What Happened Next

Research continues, but as of November 2007, most of the Universe is still unaccounted for.

Y oh Y!

It is a curse that haunts every thinking male of our species: the woeful inadequacy of the sex chromosome that defines our gender. Just look at the thing. It's a cripple, it causes trouble in all kinds of ways, and if you look at evolutionary trends, it is clearly on its way into oblivion. Rather than worrying about the size of our external sex organs, we should be worried about the decay of our Y chromosome. Geneticist Steve Jones has written an entire book about it, but I'll be brief and content myself with just this chapter.

Men may on average be taller than women, have stronger muscles and bigger cars, but at the DNA front they are clearly falling short. Instead of a second X chromosome, they only have a much smaller Y chromosome (Fig. 5). Counting only the DNA building blocks, regardless of any meaning they may or may not convey, each cell of a human male is lacking 3% of the female DNA. The loss is clearly less than compensated by the addition of 1% typical male DNA.

Researchers never had much joy with this one percent genetic masculinity. Although they discovered the factor that switches the development of an embryo towards the male characteristics (its absence produces females, even if the rest of the Y chromosome is there!) and a few genes related to male fertility, the number of genes remained surprisingly small. The majority of the Y chromosome appeared to contain evolutionary garbage, including endless repetitions and inactive copies of genes, known as pseudogenes. The only consolation for the oh-so-sensitive male pride was that of the two X chromosomes found in women, one is largely inactivated. Thus the overall difference in the amount of active genetic material isn't all that dramatic.

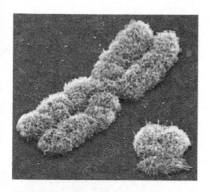

Figure 5 The human X (left) and Y chromosomes, magnified about 10,000 times. © Nature Publishing Group.

The high proportion of repetitions made sequencing the Y chromosome more challenging than the rest of the human genome. The shotgun sequencing method, which involves cutting the genetic material into random fragments, sequencing them, and letting a computer assemble the pieces in the end, has proven highly efficient in many genome projects, but fails when too many repetitions are causing confusion. The traditional method of mapping genes first and sequencing them afterwards cannot cope with this problem either.

This is why researchers from three laboratories in the Netherlands and in the US, led by David Page at MIT, had to develop a completely new strategy to tackle the highly repetitive part of the Y chromosome, which both human genome projects had put aside. They started by analyzing the Y chromosome of a single man, which they subjected to iterative cycles of mapping and sequencing. Although many had warned that the contents of the chromosome might not be worth such effort, the researchers obtained some intriguing and enlightening insights into the evolution of the sexes.

Once upon a time, a long time ago, when the evolution of mammals was just beginning, X and Y were a normal, identical pair of chromosomes. As in any of the other 22 chromosome pairs in the human genome (called "autosomes" to distinguish them from the sex chromosomes), these ancient chromosomes could exchange segments of DNA by a process known as "crossing over." This phenomenon ensures that paternal and maternal genes can be mixed in the offspring, without endangering the consistency of the contents of a given chromosome. As we shall see in a minute, the sorry state of the Y chromosome demonstrates what happens when crossing over is switched off.

Today, only a very small part of the Y chromosome still engages in crossing over with the X chromosome. This part is known as the pseudo-autosomal part of the chromosome, because X and Y behave like autosomes there. The hard-to-sequence, highly repetitive parts that have challenged researchers, however, represent the other side of the Y chromosome, and they are unable to engage in crossing over. They make up 95% of the length of the chromosome and were referred to as the non-recombining region (NRY), where "recombining" refers to processes of DNA exchange, such as crossing over. As Page and colleagues found that these regions may in fact recombine in other ways, they suggested calling them the male-specific region (MSY) instead.

The sequences which Page and his coworkers reported in 2003 reveal that the active part of the DNA contained in the MSY is a mosaic of three very different types of genetic material. Easiest to categorize are the X-transposed sequences, which contain only two of the 78 genes found in the MSY. These two have obviously been moved across from the X chromosome in a single event that appears to have happened some four million years ago, just after our ancestors parted company from the chimp's ancestors. They are the youngest immigrants in the MSY and have been found only in the human Y chromosome.

The "oldest" genes in the MSY, which means the genes that have been decoupled from crossing over with the X chromosome for the longest time and have thus had ample opportunity to evolve into typically male genetic traits, are bundled up in a second group, which Page and colleagues call X-degenerate. This group contains 16 functional genes including the gender-determining factor SRY. In addition there are 13 inactive copies of genes, referred to as pseudogenes. All 29 DNA sequences are clearly related to the corresponding parts of the X chromosome. From the degree of difference that they display now, researchers conclude that these sequences had been normal autosomal genes until the point when, some 300 million years ago, crossing over ceased and the genes started to diverge from their counterparts on the X chromosome. The different degree of divergence in different parts of the Y chromosome serves as a "molecular clock," helping researchers to draw up a timeline of how the X and Y chromosome grew more and more apart. A very slow and drawn-out divorce indeed.

The fact that, of the 29 autosomal genes that were present in this group in the early days of mammalian evolution, almost half (13) have lost their function and turned into pseudogenes highlights a serious problem inherent in the way we mammals handle our sex chromosomes. While crossing over in autosomes, XX pairs, and in the pseudo-autosomal parts of XY pairs guarantees some genetic stability, the MSY region is excluded from this stabilizing mechanism. In a way it is deeply ironic that the sex-determining part of our genome should be passed on in an asexual manner, only from father to son. This kind of inheritance carries the risk of accumulating dangerous mutations.

However, there is a way in which evolution tries to fight this risk, and researchers discovered this in the third group of DNA sequences, called the ampliconic sequences. An amplicon is a piece of DNA that is amplified, which means that multiple copies are made of it (in the lab, polymerase chain reaction or PCR is a common means to amplify a gene). And multiple copies is in fact what the MSY region does best. With more than 10 million bases, the ampliconic DNA is the largest of the three groups. It contains 9 families of active genes (i.e. 9 different genes and lots of copies and variants thereof), all of which are active mostly or even exclusively in testicles, so they can be assumed to have something to do with sex and reproduction.

Large parts of these sequences occur as palindromes, which means that a given string of DNA letters is immediately followed by a mirror-image of this string in the opposite strand, e.g.:

```
GAGATATA  |  TATATCTC
CTCTATAT  |  ATATAGAG
```

(Note that because of the base complementarity, you will find palindromes regardless of which strand you start reading.) The longest palindrome of this kind has 2.9 million bases and contains two smaller palindromes within its sequence. A large part of the Y chromosome turns out to be a rather confusing hall of mirrors!

The genes in this hall of mirrors are of different origins. Back in the 1990s, Page and others found that the loss of one group of these genes makes men infertile. The region where these genes are located is known as the azoospermia factor (AZF). Apparently, evolution has recruited these and other reproduction-related genes to the AZF region over time. This is the only known example of a chromosome specializing in a specific kind of function and "collecting" the genes related to it. In all other chromosomes, the arrangement of the genes is more or less random.

When the researchers looked more closely at the confusing multiplicity of copies and mirror images, they found a surprising solution to the problems created by the lack of crossing over, as discussed above. It now appears that the mirror-images serve as backup copies and exchange partners in a similar way as the copies found on the other chromosome in the case of autosomes. Where crossing over with the X chromosome has become impossible, the Y chromosome resorts to crossing over with itself. This self-help has been going on for millions of years, as Page and colleagues found by checking the Y chromosomes of chimps. Our closest living relatives share six of the eight large palindromes.

Researchers will need more data on many more mammalian sex chromosomes to figure out exactly how and why the X and Y drifted apart. Comparisons with species such as birds, bees, and indeed the duck-billed platypus (page 137) reveal that there are many ways of sex determination, so the crumbling wreck of a sex chromosome that we human males carry around was by no means an inevitable outcome. For now they have managed to show, however, that endless repetitions and reflections don't have to be useless and boring.

(2003)

Further Reading

H. Skaletsky *et al.*, *Nature*, 2003, **423**, 825.
S. Rozen *et al.*, *Nature*, 2003, **423**, 873.
S. Jones, *Y: The descent of men*, 2003.

What Happened Next

The year 2007 has witnessed the advent of personal genomics. Predictably, the first individuals to have their genomes sequenced were men, namely James Watson and Craig Venter. So on top of the information available on Men in general, we can look forward to lots more details on the genomes of individual (presumably great) men.

DNA Toys

The tools of molecular biology have given us the opportunity to read DNA, but they also enable scientists to make DNA and play around with it. The area of DNA nanotechnology started out as a somewhat offbeat exercise in thinking outside the box, hence its inclusion in the crazy section. A few years later, however, non-natural DNA toys became very useful DNA tools, so we shall meet this field again in a more grown-up version, in the cool section. Although DNA toys are cool, too.

The cell normally confers the tasks of building complex structures and realizing subtle chemical manipulations on specialized proteins. DNA, in comparison, may appear as a rather boring molecule, which has only four different building blocks, and whose higher order structures mainly serve the purpose of saving packing space.

Even though DNA in the cell serves a single purpose only, this does not rule out the use of its building blocks and design principles for the construction of other things. One important advantage of this biopolymer is that the tools of modern molecular biology enable scientists to handle it with unprecedented efficiency. For instance, the polymerase chain reaction provides a means of rapidly copying any DNA sequence, while restriction enzymes allow highly specific editing of the molecules.

Nadrian Seeman at New York University realized that these were ideal conditions for using DNA as a construction material. Thus, in 1990, he and his coworker Junghuei Chen set out to design and produce a nanoscale cube from DNA. Each of the six surfaces of the cube would be enclosed by a ring-shaped molecule of DNA containing 80 nucleotides, i.e. 20 on each side. Thus, each of the 12 edges of the cube

The Birds, the Bees and the Platypuses. Michael Gross
Copyright © 2008 WILEY-VCH Verlag GmbH & Co. KGaA, Weinheim
ISBN 978-3-527-32287-9

would consist of a double helix with 20 base pairs, corresponding to exactly two helical turns. At each of the eight corners, three double helices would meet to swap partners (Figure 6).

Starting from ten open-chain, single stranded DNA molecules, the researchers needed essentially five steps involving cyclization, formation of double strands, coupling, and purification, in order to obtain the final product. Using analytical methods based on the accessibility of the strands for restriction enzymes, they could prove that the construct had the right topology required for a DNA cube of their specific design. That means, the strands are wound around each other and the rings interlock exactly in the topological way required, but to prove that the structure actually had right angles and equal sides like a cube, they would have to await the formal structure determination. Which would require several orders of magnitude more material than was produced in the first demonstration – Chen and Seeman used radioactive tracer methods to analyze their DNA constructs.

This problem, however, did not stop Seeman from marching on to produce even more complex structures from DNA. With Y.W. Zhang, he published in 1994 the construction of a truncated octahedron (combine two square pyramids base-to-base, then cut all six corners) whose edges also consisted of 20 base pairs each. In total, this new DNA creation contained 1440 nucleotides, with a molecular weight of 800,000 dalton. This corresponds in size to big natural protein complexes, such as the molecular chaperone GroEL and the proteasome. This super-molecule even contains loose ends, which may in the future serve the construction of endless porous lattices similar to the inorganic zeolites.

Later on, Seeman became also interested in molecular knots like the ones discussed on page 20. Approaching this challenge his way, he also managed to get a trefoil knot from DNA, and more recently (1997) Borromean rings, that is a set of three or more rings interlocked in such a way that cutting open one of them would release all.

Apart from being fascinating toys for chemists, the DNA constructs from Seeman's laboratory may well find applications in future technologies. Although DNA is a rather expensive material, and the structural details and stability have as yet to be confirmed, the unrivalled access to specific nanoscale hollow shapes and meshworks will ensure DNA a place on the shortlist when it comes to selecting the building

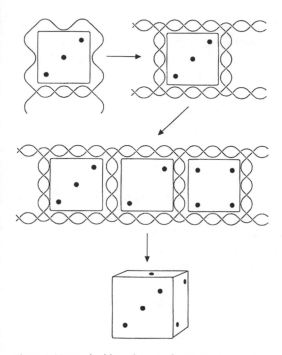

Figure 6 How to build a cube out of DNA. Each face of
the cube is defined by a ring of DNA, each edge by a
stretch where the two adjacent rings form a double helix.

blocks for the technology of the future. Possible applications could include transporter, scaffold, and even catalytic functions.

But it's not just the mechanical construction kit properties of DNA
which have aroused the interest of chemists during the 1990s. It has
also emerged that the interior of the double helix is a remarkably
efficient wire for electrons. In 1993, Jackie Barton and her group at
Caltech found that the speed of electron transfers through the aro-
matic DNA-bases stacked in the middle of the double helix like a spi-
ral staircase is extremely fast in comparison with other biological systems. The velocities she quoted were in fact so high that most people
didn't believe her. In 1995, however, Thomas Meade and Jon Kayyem,
also working at Caltech, but using different approaches, arrived at a
qualitatively similar result with slightly lower numbers, and managed
to convince the disbelievers.

In their experiment, organometallic complexes involving the heavy
metal ruthenium were used for both the emitter and the receiver of

the fast current. The emitter can be activated by a flash of laser light, and the arrival of the electron at the receiver can be conveniently observed as it changes the spectroscopic properties of the molecule. Conceivably, the complex architecture of the first ruthenium complex, which the electrons have to pass through before they reach the "electron highway" in the middle of the double helix, could be the reason why the transfer is slower than in Barton's experiment. Jackie Barton believes that her experiments, in which the electron was released directly on the "highway," measured the true speed of electron transfer in DNA.

An important outcome of these studies was the finding that the DNA "wire" only works in double helices. Even when emitter and receiver are bound to the same DNA strand, the presence of the exactly complementary second strand is absolutely essential for the electron flow. This opens up possibilities for the development of highly sensitive and specific probes for the detection of certain DNA sequences. To investigate the presence of a known DNA sequence, one could synthesize the complementary strand, couple it with the appropriate complexes, and then use the spectroscopic assay of fast electron transfer to detect the presence of the sequence of interest.

Not only can DNA replace wires, it could also one day outperform electronic computers. In November 1994, Leonard Adleman of the University of Southern California at Los Angeles reported that he had built a kind of "chemical computer" using DNA. The first problem he solved with it was a simple instance of the "traveling salesman" problem consisting in finding the shortest route linking a number of cities. Like a conventional silicon-based computer, DNA can store information in a code. Using methods of molecular biology, one can read, copy, or sort these coded datasets. Of course, each of these steps takes much longer in the DNA system than in a silicon chip. This can be compensated, however with the advantage that a reaction tube can easily contain 10^{19} different strands of DNA, representing as many different datasets. Richard Lipton of Princeton University suggested in 1995 that this massive parallelism should allow the DNA computer to solve problems that are "intractable" for conventional computers.

This prognosis has managed to wake up the computer scientists who were only moderately impressed by Adleman's molecular biology computer. Parallelism is the big issue for the future of computation, and unconventional methods to achieve it may well spell the end of the silicon chip one day.

(1995)

Further Reading

M. Gross, *Travels to the Nanoworld*, Perseus, 1999.

What Happened Next

While remaining a small field, the "misuse of DNA" has consistently come up with interesting new things in the nineties and noughties. See the cool section, pages 213 and 216, for cool and unconventional applications of DNA constructs.

Resurrecting a Billion-year-old Protein

It is quite common nowadays for geneticists to infer the genetic make-up of a common ancestor shared between two genes or even between two genomes, i.e. the last common ancestor of two separate species living today. From there, it's only one logical, but also slightly crazy step to reconstruct the common ancestor based on the inferred genetic information. At present, this would be rather tricky for complete organisms, but it can be done for individual genes, and the resulting proteins. Even if the common ancestor lived a billion years ago.

Molecular approaches to evolution and phylogenetics have mostly focused on gene sequences. From the differences between today's gene variants, researchers can infer the time when the common ancestor lived and draw evolutionary family trees with remarkable reliability.

But why stop there? If one can extrapolate back to ancient genes, one could also express those genes and produce ancient proteins to study life as it was millions of years ago. Steven Benner and his colleagues at the University of Florida at Gainesville first tried this approach to reconstruct digestive enzymes from the time when ruminants evolved an extra stomach. In 2003, he had pushed this kind of molecular revivalism even deeper into the early history of life to "resurrect" protein molecules that were last seen in the Precambrian, more than a billion years ago.

Specifically, Benner and his coworkers addressed the question of which temperature range the last common ancestor of all bacteria was adapted to. This issue has been much debated, as some of the deepest branches of the bacterial family tree lead to extremely heat-adapted species. The researchers analyzed the sequences of a helper protein in

The Birds, the Bees and the Platypuses. Michael Gross
Copyright © 2008 WILEY-VCH Verlag GmbH & Co. KGaA, Weinheim
ISBN 978-3-527-32287-9

protein biosynthesis, the elongation factor EF-Tu. As the name (Tu for temperature-unstable) suggests, this protein responds very sensitively to temperature changes and is fine-tuned to the optimal growth temperature in every organism studied. From the variations in the present-day sequences, Benner's group worked their way back to infer the sequence of the ancestral EF-Tu they evolved from.

They succeeded in expressing this billion-year-old protein in modern bacteria and studied its function at different temperatures. They also repeated the procedure based on alternative interpretations of the family tree. All the resulting ancient proteins showed temperature optima in the somewhat heat-adapted (but not extremely so) range of 55–65 °C.

(2003)

Further Reading

E. A. Gaucher et al., Nature, 2003, 425, 285.
David A. Liberles, ed., Ancestral Sequence Reconstruction, Oxford University Press, 2007.

What Happened Next

Benner and others have continued to resurrect ancient protein sequences, and analyze evolutionary relationships on that basis.

Don't Stop Me Now

Sometimes I read things in the scientific literature that sound too crazy to be true. I pinch myself; I check the cover. Yes this is really *Nature* magazine; no, it's not dated April 1st, and no I am not dreaming. One of these cases was a bunch of three papers appearing in 2004, describing how evolution has come up with highly elaborate systems to run after a runaway copying enzyme and make it stop. You just couldn't make it up. For me, of course, these crazy stories serve as a welcome excuse to go a little bit crazy myself …

It's always good to know when to stop; if, for example, I was going to carry on with this sentence, even though I have already made my point, I would go on adding more and more words becoming more and more nonsensical and you would start to think I was going crazy, and why doesn't he put a full stop, does he think he's writing *Ulysses* and so on and so forth … It would be an utter waste of time and resources.

Surprisingly, though, this is exactly what the enzyme which transcribes our genes into messenger RNA does. Every gene contains clear signals of the impending end, namely first the stop codon, a set of three "letters" that ends protein synthesis, and then the clearly marked position where the finished messenger RNA is due to be labeled with an attachment known as the poly-A tail. Nevertheless, this copying enzyme, RNA polymerase II, tends to overrun both stop signals and carry on producing useless RNA even after the proper messenger RNA has been cut off and removed for further processing. So far, researchers have found no specific piece of DNA code that would make this enzyme stop.

50   *The Birds, the Bees and the Platypuses.* Michael Gross
Copyright © 2008 WILEY-VCH Verlag GmbH & Co. KGaA, Weinheim
ISBN 978-3-527-32287-9

In 2004, several research teams reported ways in which the cell (both human and yeast cells were studied for this) can stop the polymerase from wasting nucleotide building blocks. They found evidence for something that had thus far been mooted as a crazy hypothesis, namely the torpedo mechanism. According to this hypothesis, an enzyme that digests RNA, known as an exonuclease, recognizes the dangling end of the excess RNA and starts cutting it down. (Exonucleases have to start from the outside, i.e. from a specific kind of end, while endonucleases can digest RNA from the middle.) As the exonuclease advances faster than the polymerase, the former will eventually catch up with the latter and will make it stop in an as yet unspecified way.

The group of Stephen Buratkowski at Harvard investigated this phenomenon in the baker's yeast *Saccharomyces cerevisiae*. The researchers identified the exonuclease, Rat1, and two helper proteins that link Rat1 to the polymerase so it can keep an eye on the copying machine and intervene as soon as it starts spitting out unwanted letters. Yeast can survive without Rat1 but will produce an inordinate amount of nonsense RNA (and no, it's got nothing to do with the alcohol that the yeast produces!). If the function of Rat1 is blocked by exchange of just one amino acid, the accumulation of unwanted RNA will continue, which strongly supports the hypothesis that Rat1 is the nuclease that runs after the polymerase to catch and stop it.

There appears to be a similar mechanism in human cells, but the groups of Nick Proudfoot and Alexandre Akoulitchev at the Sir William Dunn School of Pathology at Oxford discovered some additional complexities. In at least one gene they studied, the "excess" RNA produced by the polymerase forms a self-splicing ribozyme, i.e. a piece of RNA that acts like an enzyme and catalyzes its own excision from the RNA strand. (Purists will argue, however, that an enzyme or ribozyme that operates on itself violates the traditional definition of a catalyst, as it doesn't come out of the reaction unaltered. Most biochemists have decided to ignore this philosophical subtlety.)

We humans have a nuclease enzyme corresponding to the yeast Rat1, which is known as Xrn2. Alexandre Akoulitchev and his colleagues studied the cooperation between the ribozyme and this nuclease. Using detailed mutation studies, they pinned down the smallest RNA piece required for the ribozyme's self-splicing activity. They could show that the rampaging polymerase can only be stopped when

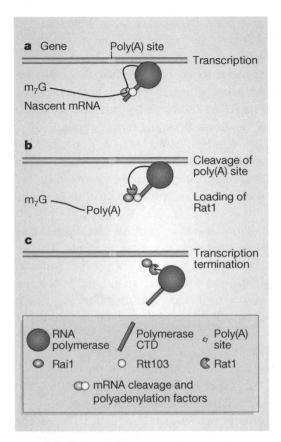

a Gene   Poly(A) site

Transcription

m₇G

Nascent mRNA

b

Cleavage of poly(A) site

m₇G

Poly(A)

Loading of Rat1

c

Transcription termination

⬤ RNA polymerase	/ Polymerase CTD ▫ Poly(A) site
◯ Rai1	◯ Rtt103 ⟨ Rat1
⬭ mRNA cleavage and polyadenylation factors	

Figure 7 Schematic view of the torpedo mechanism, believed to be used by cells when the RNA synthesis carries on beyond the end of the sequence.

(Reprinted by permission from Macmillan Publishers Ltd: Nature vol. 432, p. 456, copyright 2004)

the ribozyme is working. Obviously, the nuclease Xrn2 – unlike the yeast Rat1 – doesn't recognize the RNA produced when the proper messenger RNA is cut off, so the function of the ribozyme is to produce the kind of end that the nuclease does recognize and start digesting. And it doesn't recognize any old ribozyme cleavage site either. When the researchers replaced the ribozyme with the classic hammerhead ribozyme, the nuclease turned up its nose at the resulting RNA end as well.

So far, it is unclear how many other human genes require a ribozyme to assist the Xrn2 nuclease. As the ribozyme which the Oxford researchers discovered next to the globin gene is an atypical one, it would not have shown up in sequence searches scanning for such elements.

Now I have written up everything I know about this topic, but while I have the computer switched on and enjoy the writing, I might as well carry on speculating what would happen if my computer had an electronic version of a Ratı nuclease. As I am way past the end of the article, rambling on regardless, a little pacman style icon might appear in my text editing window, gobbling up this string of text, munching away the letters faster than I would be able to write new ones, such that after a couple of minutes it would catch up with me and I would eventually be forced to STOP.

(2005)

Further Reading

M. Kim et al., Nature, 2004, 432, S.517–522.
S. West et al, Nature, 2004, 432, S.522–525.
A. Ramadass et al., Nature, 2004, 432, S.526–530.

What Happened Next

These three papers made quite a nice splash in 2004, but I haven't heard much news from this field since then. Presumably, it takes time to elucidate the details of the mechanism and to work out how widespread it is.

Cell Jet Printers

Inkjet printers must count as one of the most miraculous inventions of our time. If only because the appliance as such, brand new, with a pair of ink cartridges, in-built scanner and photocopier, is actually cheaper to buy than a pair of replacement cartridges without the printer. Considering this madness, it appears almost sane to replace the ink with something cheaper, e.g. living cells. But will printing with cells drive scientists crazy by producing all kinds of unhelpful error messages?

Commercial inkjet printers have proven useful in the fabrication of DNA arrays and other laboratory applications. But can entire cells survive the trip through the nozzle? The group led by Thomas Boland at Clemson University, South Carolina, has succeeded in printing first bacteria and now even mammalian cells.

While printing with various kinds of molecules instead of ink has already become a routine procedure listed in textbooks, printing with cells is an entirely different matter. Depending on the type of inkjet printer chosen, cells might suffer vibrations, heat, or pressure at fatal levels. Boland's group considered both piezo-electric and thermal inkjet printers for their experiments, but found that the vibrations in the former were too powerful. In the thermal printer, temperatures can rise to around 300 °C, but the researchers hoped that the fast flow of the solution would ensure that the cells did not spend too much time in the dangerous heat zone.

Having previously shown that bacterial cells are still viable after passing through an inkjet nozzle, Boland and his group have now taken on the bigger challenge of mammalian cells, represented by Chinese hamster ovary (CHO) and rat motoneuron cells. For each cell

The Birds, the Bees and the Platypuses. Michael Gross
Copyright © 2008 WILEY-VCH Verlag GmbH & Co. KGaA, Weinheim
ISBN 978-3-527-32287-9

type, they specifically designed – in lieu of printing paper – a hydrogel material that would allow the cells to get on with their lives after the printing process. Thus equipped with printer, biological "ink," and suitable "paper," the researchers started printing. They found that more than 90% of the cells survived the process. Cultivating the printed patterns over several weeks, they observed the normal behavior expected for each cell type, e.g. the neurons made new connections with each other.

So far, the researchers have only printed the cells in a ring-shaped pattern. The next challenge will be to extend the method to biologically meaningful patterns including tissue structures and arrangements including more than one cell type. Using a four-color inkjet cartridge, a color-coded diagram on the computer screen could be directly converted into a living tissue on the gel substrate.

(2005)

Further Reading

E.A. Roth *et al.*, *Biomaterials*, 2004, **25**, 3707.
T. Xu *et al.*, *Biomaterials*, 2005, **26**, 93.

What Happened Next

Tissue engineering is an important challenge in today's bio-medical research. As I understand it, developing substrates onto which lifelike cell patterns can be printed is the main challenge. So hold on to that old inkjet printer and wait for the paper to become available.

A Frizzled Inhairitance

Topics as trivial as hairstyles practically never enter my conscious thoughts, but I made an exception for this story, partly because I seem to be carrying the frizzled gene myself. And of course the multidirectional Einsteinian hairdo is still a must for any aspiring mad scientist.

Looking a bit frizzled today? Having another bad hair day? Much as the advertisers would like us to think that the appearance of our hair is just a question of buying the right products, it may also be a question of having the right genes. As always, these hairy questions are well studied for fruit flies but not quite so well for larger animals, such as humans. The group led by Jeremy Nathans at Johns Hopkins University has now made a crucial step from flies to people, by showing that one gene from the Frizzled family controls the orientation of hair not only in the fly but also in mice.

Mice lacking this gene came up with whorls of hair in unusual places, even though they appeared otherwise healthy down to their hair follicles. Apparently it's the epithelial (i.e. skin) cells that make hairs point the wrong way. With frizzled mice brought to book, it can only be a question of months until the more hair-raising questions about the genetic origins of unruly hair in humans are answered. We'll also need to know whether there is a link to musical and artistic ability. And watch out for the next generation shampoos, "for frizzled or genetically challenged hair."

(2004)

Further Reading

N. Guo *et al.*, *Proc. Natl Acad. Sci. USA*, 2004, **101**, 9277.

56   *The Birds, the Bees and the Platypuses*. Michael Gross
Copyright © 2008 WILEY-VCH Verlag GmbH & Co. KGaA, Weinheim
ISBN 978-3-527-32287-9

What Happened Next

I haven't got a clue. As I said ...

How to Eat Without a Stomach

The mascot of life under extreme conditions, tubeworms have fascinated experts and lay readers alike. As it is very hard to breed them in captivity, research into their eccentric lifestyle has only advanced very slowly in the decades since their discovery, and surprising findings can still turn up even now.

Looking like giant paintbrushes daubed in red, tubeworms (*Riftia pachyptila*) are the most conspicuous organisms found around deep-sea hot springs and hydrothermal vents. Devoid of any digestive system, they rely on chemosynthetic bacteria living inside their "feed-bag" (trophosome) to provide them with organic nutrients. In return, the worms transport both oxygen and sulfides from their red gills to the trophosome. Surprisingly, they carry both molecules on the same transporter, an extracellular variant of hemoglobin, and manage to keep them from reacting with each other. The first ever crystal structure of such a hemoglobin molecule has now challenged the conventional view of how this transport works.

Jason Flores and his group at Pennsylvania State University solved the crystal structure of C1 hemoglobin, one of the three different extracellular oxygen carriers in *R. pachyptila*. C1-Hb has 24 protein subunits, each with the classic myoglobin fold, the characteristic arrangement of eight alpha helices uncovered by John Kendrew in the first protein structure ever solved. The subunits are arranged in two rings of three tetramers, resembling the homologue from the more mundane earthworm, *Lumbricus terrestris*. Unlike any other known globin, however, the crystal structure of the tubeworm hemoglobin showed up 12 non-heme metal ions, one in each tetramer, plus six shared by two tetramers. Mass spectrometry revealed these ions to be Zn^{2+}. This

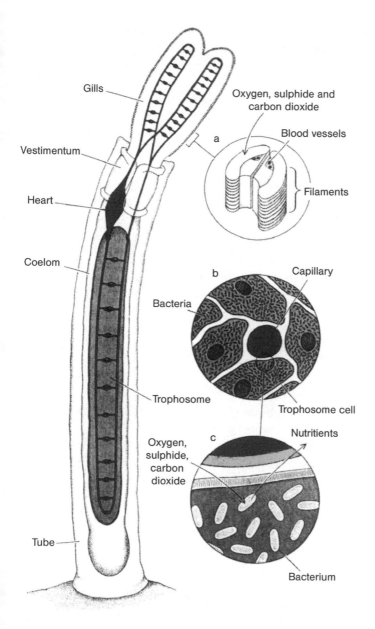

Figure 8 Cross-section of a tube worm, Riftia pachyptila, showing the trophosome, where symbiotic bacteria are housed. (Exzentriker des Lebens, 1997 © Spektrum Akademischer Verlag GmbH, Heidelberg, Spektrum Akademischer Verlag is a imprint from Springer SMB)

finding led Flores and coworkers to suspect that the zinc ions might be involved in sulfide binding, especially because the free cysteine residues, which were previously thought to bind sulfides, are deeply buried in hydrophobic parts of the structure.

Detailed binding studies revealed that the zinc ions can indeed account for most, but not all, of the sulfide binding of the protein. Flores and others report that they failed to reproduce results that were previously reported to implicate free cysteine residues in sulfide binding. Rearrangement of disulfide bridges appears to be unlikely because there have been no reports of major conformational changes associated with binding. Thus the whole truth remains to be uncovered.

(2005)

Further Reading

J.F. Flores *et al.*, *Proc. Natl. Acad. Sci. USA*, 2005, **102**, 2713.

What Happened Next

In 2007, Thomas Schweder and his team at the University of Greifswald, Germany, together with colleagues in California, managed to obtain an insight into what happens deep down inside the tubeworm, in the trophosome, where strange bacteria produce the nutrients that keep the worm going.

So far, nobody has succeeded in cultivating the bacteria from the worm's trophosome in pure culture, and maybe nobody ever will. Thus, researchers depend on the material that they can get directly out of the trophosome. Fortunately, this material turned out to be very pure; all the biomolecules it contains appear to be from a single species, namely the bacterium that feeds the worm.

In order to understand how the metabolism of the bacterium and its symbiosis with the tubeworm work, the researchers analyzed the complete set of proteins that the bacterium produces, known as its proteome. Using a classic technique of biochemical analysis, two-dimensional (2D) gel electrophoresis, they spread out the proteome into a characteristic 2D pattern of protein spots. In this method, different separation criteria are used for each dimension (e.g. mass and charge) such that each protein migrates to a specific spot in the Cartesian coordinates, where it can be identified by further analytical techniques, including sequencing.

In this way, the researchers have already identified more than 220 proteins of the inhabitant of the trophosome. Among them, they found many that were to be expected, but also a few surprises.

One fairly reasonable assumption about deep-sea symbionts, considering that they replace plants in the role of primary producers, would be that they, like plants, use the Calvin cycle to make biomolecules. Named after Melvin Calvin (1911–97), who used radioactive markers to work out the metabolic pathways of photosynthesis, this cyclic pathway funnels the energy and carbon gained from the light reaction into the production of biomolecules. One might expect chemosynthetic bacteria to use the same cycle, the only difference being that they would gain the energy from chemical reactions (such as the oxidation of reduced sulfur compounds) instead of photochemical reactions.

The investigators did in fact find all the enzymes needed for the Calvin cycle. However, one key enzyme was only found in small amounts. Rubisco (ribulose-1,5,-bisphosphate carboxylase) catalyzes the step in which carbon dioxide enters the cycle by binding to the ribulose bisphosphate sugar. In the green parts of plants, up to half the protein content can be rubisco; thus it is likely to be the most abundant protein on our planet. By contrast, Schweder and coworkers estimated that rubisco only amounts to one percent of the protein they found in the trophosomes. Therefore, they reckon that cells that have a scarcity of rubisco don't look as though they are using the Calvin cycle quite as much as plants do.

What other ways are there for the bacteria to build up biomolecules? Scrutinizing their 2D gels, the researchers found a complete toolkit for a second very famous metabolic cycle, namely the Krebs cycle (after Hans Krebs, 1900–81, who managed to get his Nobel prize-winning paper describing this cycle rejected by the journal *Nature*). In animals like ourselves, the Krebs cycle is located in the mitochondria and serves to digest biomolecules and exploit their energy. Thus it would be of little use to chemosynthetic bacteria.

However, it is possible to run the Krebs cycle backwards, and use it to create biomolecules, rather than to destroy them. Scientists have calculated that an inverse Krebs cycle might even be more energy-efficient than the Calvin cycle, given the right concentrations of enzymes and substrates. Moreover, it would explain the unusual ratio of carbon isotopes found in the chemosynthetic bacteria, which differs from the characteristic ratio in plants.

But why would the bacteria indulge in the luxury of having two complex metabolic cycles where one would suffice? Schweder and his coworkers suspect that the bacteria can switch between the two modes of metabolism according to the environment. First experimental investigations seem to suggest that when there is a lot of sulfur fuel around, they seem to prefer the less energy-efficient Calvin cycle. Only in times of need do they turn to the Krebs cycle.

Further Reading

S. Markert *et al.*, *Science*, 2007, **315**, 248.

Astronomy Helps Spotting Whale Sharks[*]

In recent years I have written quite a few pieces about research projects supported by the charity Earthwatch, which organizes the participation of fee-paying volunteers in projects around the world. While all of their projects are exciting in one way or the other, there is nothing quite as striking as the underwater astronomy project described here.

Interdisciplinary research may be all the rage, but it is still rare to find astronomical methods being used in conservation research. Yet NASA researcher Zaven Arzoumanian and marine biologist Brad Norman from Murdoch University in Australia have teamed up to do just that.

They have modified the so called Groth algorithm, which astronomers use to compare triangular patterns in photos of the night sky, to adapt it to the task of identifying spotted animals such as whale sharks (*Rhincodon typus*), the largest living fish species. Like the night sky, the whale shark sports a conspicuous spatter of bright spots on a dark background (Fig. 9).

Identifying individual animals is an important yet difficult task in conservation research. When researchers are able to identify the sharks by their spot patterns, there is no need for more time-consuming and invasive procedures, such as capturing and tagging individual animals. The researchers have shown that the "astronomical" method allows animals to be identified with equally high precision. With the help of software expert Jason Holmberg, they have already incorporated the new method into an online photoidentification library

* First published: M. Gross, Current Biology 2006, 16, No 1, R3–R4 Stellar insights

(http://photoid.whaleshark.org), which archives digital images of whale sharks.

Brad Norman is enthusiastic about the performance of his new research tool. "The program even made matches that I had initially missed and it has since made more than 100 new matches that were later verified manually. We've now eliminated the need to scatter all the images on the floor."

With the help of the environmental research charity Earthwatch, Norman is now able to apply the technique on a large scale, and literally anybody can join in. Earthwatch has pioneered new ways of involving fee-paying volunteers in research projects that would otherwise be unfeasible owing to the lack of manpower. Founded in 1971 in Boston, Massachusetts, Earthwatch now employs 3,500 volunteers per year who work on 140 different projects around the globe.

Brad Norman's Earthwatch-sponsored project, "Whale Sharks of Ningaloo Reef," is aimed at identifying how tourism and ocean conditions are affecting the survival and migration behavior of whale sharks, which feed by filtering sea water and are entirely harmless. Between April and June 2006, eight teams of six to eight volunteers worked one week each to observe and photograph the animals in Ningaloo Marine Park, off the coast of Western Australia, where these and other ocean dwellers have become a focus of flourishing ecotourism.

Specifically, the research is designed to establish the rate of return of the animals to the region and to specific parts of the reef, any site preferences of individuals, their growth rates, and their behavior under the pressure of a growing tourism industry (although this is wide-

Figure 9 Whale sharks are peaceful animals. Their characteristic pattern of bright spots on a dark background allows researchers to identify individual animals. (Brad Norman/Earthwatch)

ly applauded as an example of "best-practice" ecotourism). So far, conservation efforts for this threatened species are hampered by the lack of the most essential information, including global population size and typical life histories of individuals. By manning the observation project with volunteers throughout the shark season and over several years, Norman hopes to expand knowledge in these areas and thus help the efforts to save the species.

Although the methodology was developed specifically for the whale shark project, the researchers emphasize that it is widely applicable to other species as well, as long as they have spots. Roger Mitchell, chief scientist at Earthwatch Institute (Europe) points out that "lions, for example, have small dark spots where their whiskers attach to their noses, and this differentiating feature might help to track them in the near future."

I'm wondering whether whale sharks ever look up at the night sky, and what they make of it ...

(2006)

Further Reading

Z. Arzoumanian *et al.*, *J. Appl. Ecol.*, 2005, 42, 999.
www.earthwatch.org

What Happened Next

Pattern recognition software has been all the rage in conservation research in recent years. In 2007, the UK National Centre for the Replacement, Refinement, and Reduction of Animals in Research, (NC3Rs) announced funding for a range of projects including one that uses a similar approach to identify individual frogs in the lab. The fundamental conundrum that Matt Guille at the University of Portsmouth is going to address with his research is that frogs need to be kept in large groups so they can live their lives normally, but that they still have to be identified individually. While all existing methods of identification are invasive or harmful to some extent, Guille hopes to develop a method "which measures the pattern on the backs and feet of the animals using digital imaging and therefore is not harmful to the frog. If successful, this technique will be marketed commercially."

Talk to Your Proteins

Way back in 1996, I published a theoretical paper with the title "Linguistic analysis of protein folding." When I wrote it, I could not find any other paper combining the key words of linguistics and protein, but I found out later that somebody else published another one just a month or so before mine appeared. After that, this crazily interdisciplinary field remained unexplored for many years, but recently other researchers have returned to it, giving me a chance to blow my own trumpet and cite my own paper.

Letters, letters everywhere. As the genome sequences keep flooding in and scientists derive thousands of new protein sequences from them, the need for new methods to make sense of all this information overload is more urgent than ever. Italian protein researcher Mario Gimona suggests that approaches borrowed from linguistics might revolutionize the analysis and annotation of data concerning the proteome, i.e. the entirety of the proteins coded for by the genome of a given species.

While linguistic analysis of genes has been performed for decades, similar considerations for proteins only started to emerge in 1996. The analogy starts out simply enough, as both protein sequences and English sentences are made up of "letters" representing the 20 amino acids in one case and various sounds in the other. The challenge is to work out how to move up to the next levels of complexity, to be able to read, the "words," "phrases," and "sentences" of the protein language. "This problem is closely related to the as yet unresolved 'prediction problem' of protein folding," says folding expert Kevin Plaxco from the University of California at Santa Barbara. "Ideally, we would want

The Birds, the Bees and the Platypuses. Michael Gross
Copyright © 2008 WILEY-VCH Verlag GmbH & Co. KGaA, Weinheim
ISBN 978-3-527-32287-9

to be able to read any string of amino acids and understand its 'meaning' as easily as we read an English sentence."

For this to succeed, one would need to understand the "grammar" of protein structure and folding. While a coherent protein grammar covering all levels from amino acid sequence through to functional interactions has not yet been proposed, Gimona suggests that protein modules, i.e. the independently folding domains that can serve as intermediate scale design elements in larger proteins and have often been duplicated and re-used by evolution, hold a key position for the understanding of protein grammar. "Domains and modules represent the syntactic and semantic units in a protein," he suggests. He is optimistic about the application of linguistic methods in protein science: "When the smoke has cleared, we all might have become molecular linguists!"

(2006)

Further Reading

M. Gross, *FEBS Lett.* 1996, **390**, 249.
M. Gimona, *Nature Rev. Mol. Cell Biol.* 2006, **7**, 68.

What Happened Next

I'm still hoping that the use of linguistic methods in protein structural biology breaks through big time, if only so that more people get to cite that seminal paper of mine ...

Ancient Stowaway in Our Eyes

For a few years, I took pride in the title of the science writer in residence at the department of crystallography, Birkbeck College, London. One of the research fields pursued at this department is the study of eye lens proteins, their structures and evolution. The following story is about a structural protein of the vertebrate eye lens which has been identified as a close relative of the bacterial glutamine synthetases, suggesting that "moonlighting" saved the protein from oblivion when its original role was taken over by a different family of enzymes.

Eye lens proteins have kept surprising scientists both with their life-long stability and with their unusual evolutionary history. Now researchers have found a bacterial enzyme, which in higher organisms lost its original function to a different family of proteins, surviving in a structural role in the vertebrate eye lens.

Graeme Wistow and his coworkers at the National Eye Institute in Bethesda, Maryland, US, together with researchers at Birkbeck College London, UK, have carried out genetic and structural investigations on the protein lengsin (for lens glutamin-synthetase-like protein) from mouse and other vertebrate species.

Both the sequence comparison and the structural analysis by cryo electron microscopy pointed towards a surprising evolutionary history. Unlike the vertebrate glutamine synthetases, lengsin assembles to a symmetrical complex of 12 subunits, as does the ancient bacterial version of the enzyme. Similarly, reconstruction of family trees with the help of mutation studies revealed the vertebrate lens protein to be a member of the bacterial family of glutamine synthetase proteins known as GS I.

The Birds, the Bees and the Platypuses. Michael Gross
Copyright © 2008 WILEY-VCH Verlag GmbH & Co. KGaA, Weinheim
ISBN 978-3-527-32287-9

It was the first non-bacterial member of the family to be discovered, but a subsequent search of the sea urchin genome revealed several candidate genes for similar proteins in this organism. So far, it is not known where the sea urchin – which has no eyes – expresses the corresponding proteins and what they are used for.

Wistow and colleagues tested the lengsin from the mouse eye lens for any enzymatic activity related to the ancestral glutamine synthetase function and found none. In agreement with this finding, one of the amino acid residues considered crucial for glutamine synthetase is found mutated in vertebrate lengsins. The researchers suspect that it may serve a structural role in the eye lens.

There are previous examples of enzyme proteins being used for different functions in the eye lens, as Wistow explains: "We are used to the idea of enzymes, such as lactate dehydrogenase, being recruited to serve new roles in the vertebrate eye lens."

However, in this case, there is a new twist to the evolutionary tale. "A big surprise in this story was the realization that lengsin is very old and indeed comes from a family of enzymes that was known only in prokaryotes," Wistow said. "It seems that enzymes of this family were actually expressed in very ancient ancestors of vertebrates but were lost during evolution. One member survived in vertebrate genomes because it acquired a different, specialized role in the lens, losing its ancestral catalytic role in the process," he added.

(2006)

Further Reading

K. Wyatt et al., Structure, 2006, **14**, doi: 10.1016.j.str.2006.10.008

Deciphering the Secrets of the Neanderthals

In 2001, the human genome was more or less complete, after a huge international effort taking up the best part of a decade. Only five years later, researchers set their sights on the genome of the extinct hominid species the Neanderthal. Need I remind you that there are more than six billion specimens of *Homo sapiens* running around on our planet, who could provide tons of human DNA if that much were needed, while the Neanderthal project started out with a corroded, fragmented, and generally rotten DNA sample weighing all of 50 milligrams. Now if that isn't crazy, I don't know what is. However, by the time I had read the papers, the researchers had convinced me that this was a crazy project worth undertaking. Plus I've always had a soft spot for Neanderthals and have even done some digging myself.

Cracking Old Bones

Genome research on extinct species has often been ridiculed and assigned to the fictional realm of *Jurassic Park*. Only very few fossil bones are sufficiently well preserved to give researchers hope they might retrieve the DNA of the original owners. And even if DNA can be found, it is highly fragmented and mixed with the genetic material of other species, such as soil bacteria.

These fundamental problems haven't gone away. But what has changed in the last few years is that the methods of genome research are now so advanced that they can cope with this kind of situation. In modern genome sequencing, according to the shotgun approach pioneered by Craig Venter, the DNA will be blown to bits anyway, and the typical fragment size of ancient DNA is now well within the range that

The Birds, the Bees and the Platypuses. Michael Gross
Copyright © 2008 WILEY-VCH Verlag GmbH & Co. KGaA, Weinheim
ISBN 978-3-527-32287-9

Figure 10 The Neanderthals, *Homo neanderthalensis*, became extinct some 30,000 years ago. Their genome information is useful not only for historical questions, but also to help scientists understand genetic diversity in present-day humans. © Neanderthal Museum / M. Pietrek.

sequencers can use routinely. As for the contamination with DNA of other species, the availability of hundreds of bacterial genomes in databases – along with the human one and dozens of other genomes of multicellular organisms – allows researchers to check every fragment that they sequence and to verify what type of living being it comes from.

Basing their work on such considerations, Svante Pääbo and his coworkers at the Max Planck Institute for Evolutionary Anthropology at Leipzig, Germany, set out to investigate the nuclear DNA (i.e. the bulk of the genetic material stored in the nucleus, as opposed to the minute amount of DNA contained in mitochondria) from bones of our closest (though extinct) relative, the Neanderthal. In this case, the most dangerous kind of contamination is DNA rubbed off from the fingers of the people who handled the bones, as modern human DNA is much more similar to Neanderthal DNA than that of any other species might be.

In order to identify samples with minimal contamination by *Homo sapiens* DNA, Pääbo first looked at the mitochondrial DNA of Neanderthals, which had already been studied in great detail. Starting out with 70 bones and teeth from various Neanderthal sites, the researchers first identified six samples in which biomolecules were relatively well preserved. Using a polymerase chain reaction (PCR), they then went on to analyze the relative amounts of mitochondrial DNA from Neanderthals and from modern humans in each of these samples. While in most of them, the overwhelming majority of the DNA found was modern, the researchers were able to identify a single bone – which had been found in the Vindija cave in Croatia, and is therefore called Vi80 – in which the primate mitochondrial DNA was almost pure Neanderthal material. (Note that this DNA sample still contains vast amounts of bacterial and possibly fungal DNA, but these are easy to discard during sequencing.)

For the analysis of this precious bone, the Leipzig team used a brand new method that not only speeds up the process by two orders of magnitude but also enables them to keep an eye on each individual fragment, so they will know, for example, which of the two strands of the double helix a given fragment represents.

Developed by Jonathan Rothberg and his colleagues at 454 Life Sciences in Connecticut, this new method is based on the incorporation of a certain base during the synthesis of a complementary strand and on the release of pyrophosphate (which is why it is known as pyrosequencing – no pyrotechnics here). Rothberg's team developed a complete integrated system which can carry out this kind of sequencing on immobilized single strands of DNA in reaction wells of picoliter (10^{-12} l) volume.

The method works best with fragments of around 100 bases in length. In a single session lasting no more than four hours, the automatic system can sequence 25 million bases, as Rothberg demonstrated by sequencing *Mycoplasma genitalium* all over again. Back in 1995, this bacterium had been the second species ever to get its entire genome sequenced.

This new sequencing method arrived at just the right time for Pääbo's project. The bottom line is that it works 100 times faster and even more precisely than the classical Sanger method that is now routinely used, with capillary electrophoresis and fluorescent dyes taking care of the separation and detection, respectively. The disadvantage of pyrosequencing is that it doesn't perform so well on fragments much longer than 100 bases. For paleo-genome investigators such as Pääbo, this point is irrelevant, as their DNA samples are already shredded to small pieces when they arrive in the lab and rarely amount to more than 100 base pairs in length.

Thus, the Leipzig research team could very happily unleash the newest sequencing technique to crack the secrets of the old bones. In the first part of their project, they sequenced around a quarter of a million different DNA fragments from the Vi-80 bone and tried to assign these to a large group of organisms by comparison with genome sequences in existing databases. For around 200,000 sequences, the attempt failed. Of the 50,000 that could be categorized, 17,000 turned out to be from soil bacteria, from the group known as Actinomycales.

The second most prominent group, however, was primate DNA, comprising 15,701 putative Neanderthal sequence fragments. Among these, the researchers first picked out the 41 mitochondrial DNA sequences in order to reassure themselves (again) that the primate in question wasn't a mundane *Homo sapiens*. All 41 passed this test. Moreover, the analysis of these fragments proved useful in expanding the existing knowledge on Neanderthal's mitochondrial DNA and delivered an estimate of the time when the two hominid species parted company. According to these samples, the separation happened between 461,000 and 825,000 years ago.

With this reassurance, the researchers turned towards the main goal of their investigation, namely the as yet unexplored DNA contained in the nuclei of the Neanderthalian cells. In November 2006, Pääbo's group reported the first one million base pairs of the Neanderthal genome, corresponding to some 0.036% of the whole genome. The yield was a little lower for the sex chromosomes, but as both X and Y fragments were identified, it is certain that Vi-80 belonged to a male Neanderthal. The first million was, of course, only a randomly assigned benchmark, and the researchers have since then continued to sequence much more material.

Based on the first million base pairs, the Leipzig researchers have already been able to carry out detailed comparisons with the known genome sequences of our own species and of the common chimpanzee, *Pan troglodytes* (see page 141). Of course, the overwhelming majority of DNA bases is identical in all three species.

Using the chimp genome as an external standard, the researchers were able to attribute differences between modern humans and Neanderthals to the evolutionary change of one or the other. Our genetic material, for instance, differs in 434 positions (out of the one million studied) from the consensus between Neanderthal and chimp. Assuming that mutation rates of *Homo sapiens* and *Homo neanderthalensis* were similar, there should be a similar number of differences between Neanderthals on the one hand and the chimp/human consensus on the other.

What the researchers found, however, was that the Neanderthal DNA appeared to have eight times as many specific mutations as modern human DNA. Their interpretation is that roughly seven out of eight variations found exclusively in Neanderthal DNA may be due to damage that the ancient DNA suffered during the tens of thousands of years it spent in the ground. Until the day when DNA from other Neanderthal individuals becomes available for comparison, scientists have no choice but to disregard these presumed mutations, and to focus on the specific differences exclusive to the modern human genome, which they can easily check as many times as they like.

Based on the sequence comparisons between these three cousins, Pääbo and colleagues have estimated the time of separation between Neanderthals and modern humans to be around 516,000 years before our time. They have pointed out, however, that this figure is very sensitive to any errors in the separation date between us and chimps, presumed to be 6.5 million years ago, so it should not be considered a final truth.

The preliminary study of the first million base pairs also allowed insights into the size of the founding populations. Unlike other primates, both Neanderthals and *Homo sapiens* appear to be descended from relatively small populations of around 10,000 individuals. The Neanderthal data is also useful for researchers interested in the origins of human genetic diversity. There are many positions in the human genome where no consensus version seems to exist, but several variants are found to be widespread. If these variations only involve a single base pair, they are referred to as SNPs (single nucleotide polymorphisms, pronounced: "snips"). With Neanderthal and chimp as reference points, it is now possible to find out which version of a given human SNP is the "original" one and which the mutation. In some cases, of course, the variability may have originated even before the split from Neanderthals.

Almost simultaneously with Pääbo's paper in *Nature*, there appeared a second Neanderthal genomics study in the competing journal, *Science*. Edward Rubin and coworkers in several research institutes across the US analyzed DNA from the very same bone with a somewhat different approach. They only sequenced 62,500 base pairs, but selected these beforehand, so they may have obtained a comparable amount of meaningful data to the Leipzig researchers who looked at random DNA fragments. Rubin's preliminary conclusions diverged from Pääbo's in some key aspects, such as the dating of the split and the possibility of genetic mixing between the two species.

Based on these preliminary results and the proven feasibility of Neanderthal genomics, Pääbo and colleagues are now firmly convinced that the sequencing of the entire genome is within their reach. Whole genome comparisons between ourselves and our Neanderthal and chimp cousins, they argue, would yield unprecedented insights into human evolution and genetic diversity. Beyond the analysis of simple differences such as SNPs, comparisons would also include other kinds of genetic diversity, including copy number variations, whose importance scientists have only begun to appreciate quite recently.

Regarding the rise and fall of our closest cousins, there is an ongoing debate on whether Neanderthals and early humans had any contact after the species separated, whether any gene exchange took place, and whether humans had any active part in the extinction of Neanderthals. All these fundamental questions will be easier to assess with the help of full genome comparisons. Pääbo estimates that this dream could become reality as early as 2008.

(2006)

Further Reading

M. Krings et al., Nature Genet., 2000, 26, 144.
M. Margulies et al., Nature, 2005, 437, 376.
R.E. Green et al., Nature, 2006, 444, 330. (one million base pairs)
J.P. Noonan et al., Science, 2006, 314, 1113. (62,500 base pairs)
R. Redon et al., Nature, 2006, 444, 444.

What Happened Next

In September 2007, controversy flared up regarding the possibility of human contamination in Pääbo's samples. Further controls may be needed to clarify how much of the original data can count as truly 100% Neanderthal. Meanwhile, the proposed sequencing of the entire genome by 2008 may get delayed by these problems.

In an unrelated paper published in the same month, innovative use of carbon dating allowed researchers to cross-check proposed extinction points of Neanderthals directly against climatic events (instead of linking both to a calendar, which is fraught with uncertainties). It appears from the new analysis that there is no clear link between climate change and the demise of the Neanderthals.

Oh, and the sequencing technology mentioned in this piece is the same that served to complete one of the first personal genomes, that of James Watson. I'm sure there is a joke in this somewhere, but I'll leave that to you.

Meet the Family

Homo neanderthalensis was our closest relative in the family tree of the hominids. Around half a million years ago, Neanderthals emerged as a species separate from ours when they moved into Europe and adapted to the colder climate, while our ancestors remained in Africa for the time being. Thus, the evolutionary distance between these two species is roughly a tenth of the distance separating us from chimps.

The name comes from the Neandertal valley near Düsseldorf, Germany, where a specimen was found some 150 years ago. Unfortunately, however, this first Neanderthal skeleton had a bit of a rough ride. The cave where Mr Neanderthal used to live was destroyed during quarrying, and the first bones were discovered in the rubble. Not until the 1990s did researchers succeed in tracking down the remainder of the material and rescuing additional bones.

The original find included the skull, which after a few years was classified as a new species owing to the characteristic bulges above the eyes and the receding forehead. Since then, numerous specimens have been found across Europe and the Middle East. Neanderthal skeletons often occur together with a certain kind of stone tools, known as Mousterian (after the site Le Moustier, Dordogne, France).

Around 40,000 years ago, modern humans expanded from Africa into Europe. Some 30,000 years ago, Neanderthals became extinct, while our ancestors, represented by the Cro-Magnon find among others, appeared to flourish. The big challenge for hominid anthropology is to figure out what happened in those 10,000 years. Did the Neanderthals retreat to escape from the technologically more advanced competitors? Might our ancestors have killed them off more directly? How long did the co-existence in Europe last, and was there any contact or even interbreeding?

In 2006, Mousterian tools were discovered in a cave in Gibraltar, dated to 28,000 years before our time. This finding suggests that the last survivors of the species found refuge in the mountains of Southern Spain, and that the co-existence may have lasted longer than scientists had previously thought.

Even the hypothesis – discounted by most experts – that the 24,500 year old skeleton of a child from Lagar Velho (Portugal) may be a cross-breed between Neanderthal and modern human has gained a little more credibility after the Gibraltar discovery. The story of our closest relatives remains an intriguing mystery.

Further Reading

C. Finlayson *et al.*, *Nature*, 2006, **443**, 850.

How to Find Neanderthal Remains

On a hot July afternoon, my daughter (then 12) and I reluctantly climbed out of an air-conditioned train onto a sizzling platform, apparently stranded in the middle of nowhere, somewhere in Southern Spain. Only when the train left could we see that there was actually a minuscule station on the other side of the tracks, and that a paleoanthropologist – easily identified by a picture of hominid skulls on his tee-shirt – was waiting to meet us. We had come to this remote area in order to help as volunteers in an excavation at a cave site where dozens of Neanderthal remains had been found over the previous ten years. No previous experience required, four meals and youth-hostel style accommodation included in the price. The adventure starts here.

The Tuesday of our arrival is taken up by the changing of the guard. Another 15-strong team of volunteers has worked in a different cave further north for three weeks, and they have come down to see "our" cave before they leave. They are in high spirits, having discovered a hand-axe at their site, which made the news on national television. At lunch time, we are the only newbies at the table, but by dinner time the leavers have left and most arrivals have shown up. Wednesday should be a working day, but the professor is kept away by other duties, so we only do a guided tour of the excavation site and a few other caves in the same mountain. The cave is essentially a deep vertical hole in the limestone rocks, big enough to house a builder's scaffold.

It is accessible through a horizontal tunnel dug by miners or through the upper end.

As we, unlike some, actually enjoy the climb up to the entrance of the cave, we get the privilege of being part of the first team of three diggers to work on the actual excavation site, which is operated from the top of a scaffold, just below the entrance. Materials go down the cave by pulley and are then carried out through the miners' access. At the "upper cutting," we have two excavation fields, measuring a square meter each, which are taken off very slowly with small trowels. The relatively loose sediment is peppered with bones, but sadly, most of them stem from small animals, such as rabbits. Bits and pieces of deer or turtle are much rarer. No signs of humans or other big carnivores on the first day. Bones picked up during the excavation work go into a tray, and all the rest goes down the pulley in buckets, to be collected in bags at the bottom.

As the heat makes it impossible to work at the site after 2 pm, there is an extended lunch break including a visit to the swimming pool and time for a siesta. In the late afternoon, we go back to work, taking the bags from the excavation site to the next step, namely the sieving. This takes place at the marble factory on the other side of the mountain, where there is a hose with a particularly strong water jet, which helps to break up the dirt and pass it through the set of three sieves with decreasing mesh sizes. Then we scrutinize the sieves and pick out all bones, anything that might be a stone tool, and even snail shells. The latter are of interest not for their own sake, but for the samples of soil they encapsulate. All these finds go into trays, labeled with the date and location of origin, but so far unsorted.

Friday is the day of the local patron saint, so the marble factory is closed and there is no wet sieving. Instead we install a huge dry sieve at the miners' entrance to the cave and sieve through the rubble dumped there. Such second-hand material is less valuable because it lacks precise geological context, but then again, one can process it more rapidly and find more and bigger bones with less trouble. And we shouldn't forget that major finds have been made in other people's rubbish. At the Neanderthal site, where quarry workers found the original skull cap that led to the definition of the species *Homo neanderthalensis*, the recent re-investigation of the soil that those workers discarded yielded a wealth of Neanderthal remains. Oh well, one is allowed to dream, isn't one. In this particular heap of rubble we find pieces of antlers, turtle shells, and a flint flake, but no human bones.

Late Friday afternoon we get to know the last step of the procedures, sorting the finds. There are more than a dozen different categories, for bones and teeth that are burnt or unburnt, from small or large animals, classifiable or unclassifiable. Unclassifiable is a popular choice, as all bones in that heap – typically pieces of bone shaft without a joint or a specific shape that would allow identifying their origins – will go in one plastic bag, never to be looked at again. In contrast, each bone deemed to be classifiable has to be put into a small plastic bag of its own, with a label. Our task ends with the bagging, counting, and combining bags into bigger bags. But some unfortunate biologists will one day face the task of looking at all the classifiable bones again and work out the stories they may be able to tell us. It appears that the cave sheltered not only Neanderthals but also some other meat eaters, such as foxes or owls. From the relative abundance of the various bones (vertebrae are suspiciously scarce), researchers may be able to figure out who ate whom and when.

After the sorting exercise, the Professor holds a talk about the previous work at the two sites. He shows some casts of the human remains found previously, and the original hand-axe discovered a couple of weeks ago. Seeing the hand-axe, a vaguely oval, fist-sized piece of rock with bits chipped off only around the edges, and the sharp end broken off, comes as a shock to most of us. Imagine that thing turning up caked in dirt – I am sure we wouldn't have recognized it as a tool (as in fact the person who dug it out didn't, either). For all I know, I may have thrown five hand-axes away during the dry-sieving. Oh well. I'm sure someone else will find them when they sieve through our rubble.

Saturday we are offered a tour of the other site, plus a Roman monument and a recently discovered Neolithic round house. Sunday, the marble factory is still closed, so it is another day of dry sieving and sorting. No hand-axes in sight, even though we now know what they are supposed to look like.

Monday is our last working day, as we have only booked in for the first week of the three-week dig. First thing in the morning we have another go at the excavation fields together with a newly arrived volunteer. He gets lucky and finds a femur head, which is the ball-shaped end of the thigh bone, which goes into the hip joint. It might be human, but then again, it might be any other large mammal just as well. Biologists will have to have a closer look at that one.

In the afternoon, we're back at the marble factory scrutinizing the wet sieves. It's getting late, and as we are beginning to tackle the very last sieve, my daughter dips her hand in and pulls out something that was a bit whiter than the rocks and dirt. "This could be a human tooth, couldn't it?" she asks me. As an absolute layperson in anatomy, I think it could, but don't trust my judgment, so I turn to the assistant standing next to me and ask him the same question. He looks at me as though he is absolutely certain I am trying to pull his leg, then he looks at the tooth and changes his expression. "In fact, yes, it is a human tooth." Of course we examine every speck of dirt in this lucky sieve very closely, but to the end of the first week, this tooth remains the only unequivocally Neanderthal find. On Tuesday, we're off to the beach, but the rest of the people have two more weeks to find further remains of the owner of our tooth.

(2003)

Eat Isotopes and Live Longer

Quite often, exciting stories find me before I find them. The researcher behind the following story approached me and persuaded me to do something about it. I hesitated at first as I am not entirely convinced that it's a good idea to make rich people live longer while poor people keep dying prematurely of entirely avoidable causes. However, as the science behind this appeared sound and the topic certainly had the potential to interest a wider audience, I eventually took it on and wrote a news item including some caveats. A less inhibited colleague simultaneously published a more enthusiastic news piece for a different magazine, accompanied by a press release which was picked up by mainstream media around the world. So the story made quite a nice splash, which is always fun to watch.

An Oxford researcher has suggested that heavier but stable isotopes such as deuterium and carbon-13 could be used to suppress the ageing reactions attributed to reactive oxygen species.

Reactive oxygen species (ROS) are a staple of ageing research, as they are believed to cause cumulative damage to the molecular inventory of the cell. Oxford-based researcher Mikhail Shchepinov has now suggested that food containing heavy isotopes of hydrogen, carbon, and nitrogen in key positions could drastically reduce oxidative damage or even avert it altogether.

Shchepinov's argument is based on the chemistry of the processes by which biomolecules such as DNA, proteins, and lipids suffer oxidation damage in the cell. Typically, the rate-limiting step of these reactions is the removal of hydrogen from the carbon atom to be oxidized. If the carbon and/or the hydrogen involved are replaced by a

The Birds, the Bees and the Platypuses. Michael Gross
Copyright © 2008 WILEY-VCH Verlag GmbH & Co. KGaA, Weinheim
ISBN 978-3-527-32287-9

heavier version of the same element (e.g. carbon-13 or deuterium), the reaction will be slowed down. This well-established phenomenon is known as the kinetic isotope effect.

How does one get these isotopes to the crucial sites in the cell? It is easiest when the molecular building blocks in question are essential nutrients, which means that the body cannot synthesize them from scratch and they have to be taken in as food. Several amino acids that are known targets of oxidative damage in proteins are also essential nutrients, which suggests that they can be protected with the help of the isotope effect. The situation is similar for lipids (derived from essential fatty acids).

Nucleic acids are a more difficult target, however, as their building blocks can be synthesized in the body. However, Shchepinov argues that these molecules are conditionally essential, which means that in certain situations (e.g. after fasting) they can become essential.

Shchepinov, who has already applied for several patents on this approach, is confident that this is the way towards a longer and healthier human life span. "The first biological experiments conducted in Russia were very promising," he said, although he did not want to divulge any details as yet.

The known fact that heavy water is toxic to higher organisms does not faze him either. "The isotopes will only be incorporated in the sites that need to be protected from oxidation," he argues. "Ideally, they will slow down the oxidation reaction so much that they will never be released to take part in other reactions. If some of them do break free, they will only occur in extremely small concentrations."

Turning this idea into a commercially viable practice will be a different matter, though. As the modified amino acids will have to be synthesized rather than just grown in a field, isotope-enriched food will certainly be much more expensive than what we eat now. To have any effect on aging, the special food would have to be on the menu every day, for life, which may end up being affordable only for the mega-rich. Of which there are a few, admittedly, but as a society, we might not want to widen the gap of life expectancy between rich and poor even further.

Even if the isotope food became more affordable, it may be hard to sell it to a public already concerned about chemical additives, GM crops, and any other tampering with food. Thus, only the Mr Burnses of this world, with unlimited cash and an unhealthy interest in immortality, may end up eating their way into an isotope-sponsored second century of life, while the rest of the world may not even notice.

Conceivably, though, there will be other niche markets for isotope food. Long-distance space travel springs to mind. As astronauts eat synthetic food anyway, and are exposed to increased oxidation damage due to radiation, it is to be hoped that isotope food will be available in time for the first astronauts traveling to Mars.

(2007)

Further Reading

M. Shchepinov, *Rejuvenation Res.*, 2007, **10**, 47.

Virulence From The Deep Sea

When I wrote my book "Life on the Edge" (about life under extreme conditions) and included a section on the stomach bug, *Helicobacter pylori* (see page 6), I felt I was pushing the envelope. Yes, *H. pylori* is adapted to an extremely hostile environment, the human stomach, but as this environment is quite distant from the other biotopes discussed in the book, such as deep-sea hot springs, and Antarctica, these different kinds of extremism must be entirely unrelated? Ten years later, it turned out that there is in fact a surprising link between the eccentric bugs that cause stomach ulcers, and those that live peacefully at the bottom of the ocean...

The chemosynthetic bacteria that feed the lightless deep-sea biotopes have evolved crucial traits that serve as virulence factors in their distant, pathogenic relatives, as the first analysis of their genomes has uncovered.

The biotopes surrounding hydrothermal vents and hot springs in the deep sea depend on chemosynthetic bacteria, which often live in symbiosis with the resident animals such as clams and tubeworms (see page 58). Genome sequences of two of these symbionts have revealed surprising similarities with common human pathogens including *Helicobacter* (stomach ulcers) and *Campylobacter* (food poisoning) species.

Until very recently, the deep-sea endosymbionts resisted all attempts to grow them in pure cultures, which would normally be the very first thing that microbiologists do before they set out to characterize and name a new species. Researchers had to resort to studying

The Birds, the Bees and the Platypuses. Michael Gross
Copyright © 2008 WILEY-VCH Verlag GmbH & Co. KGaA, Weinheim
ISBN 978-3-527-32287-9

crude samples, e.g. directly from the trophosome of the tube worms (see page 60).

The group led by Satoshi Nakagawa at the Extremobiosphere Research Center at Yokosuka, Japan, recently managed to obtain pure cultures of several strains of epsilon-proteobacteria from hydrothermal vent communities, which they strongly believe to be symbionts (even though their remote-control sampling technique doesn't allow them to pin down exactly in which environment the bacteria live). Now the same group has completed the genome sequences of two such strains, which they recognized as members of the genera of *Sulfurovum* and *Nitratiruptor*, respectively.

The genome sequences, analyzed in comparison to the substantial body of sequences both from close and not-so-close relatives, revealed many details of the adaptations that allow these bacteria to thrive in extreme physical conditions and in the absence of photosynthesis. Like the bacteria from tubeworms reported earlier in 2007, these two species have the complete assortment of genes necessary to run the Krebs cycle backwards. This is a key set of reactions in higher organisms, which the bacteria use in order to create organic molecules, rather than to digest them (as we do in our Krebs cycle).

More surprisingly, however, the researchers also obtained valuable insights into the adaptation of bacteria in a very different extreme environment: the human stomach. Nakagawa and his colleagues were surprised to find several genes believed to be crucial virulence factors for the human pathogens *Helicobacter* and *Campylobacter*, which can survive in the highly acidic environment of our stomachs. The authors conclude that the "deep sea vent chemoautotrophy has provided the core of virulence for important human and animal pathogens."

It is quite common for pathogenic bacteria to pick up genetic traits from other species, in a process known as horizontal gene transfer. For instance, the dreaded hospital superbug MRSA is a strain of *Staphylococcus aureus* that has acquired an additional gene, making it resistant to antibiotics such as methicillin. Often, such traits reside on circular pieces of DNA called plasmids, which can easily be transferred even between bacteria of different species. Even if the transfer of one specific gene to a given recipient may be a rare event, the presence of a matching evolutionary constraint, such as an antibiotic, will give the recipient an advantage, and the trait can spread rapidly.

The virulence factors shared between human pathogens and deep-sea bacteria may have crossed over from one to the other in similar ways, but this requires at least that their distant ancestors have lived closer to each other. Where exactly the species met and exchanged weaponry remains an intriguing mystery.

(2007)

Further Reading

S. Nakagawa *et al.*, *Proc. Natl Acad. Sci. USA*, 2007, **104**, 12146.

2

Sexy Science

"Gravitation is not responsible for people falling in love".

Albert Einstein

I write about molecules more often than about people, which is a little unwise economically speaking, as there are so many more readers for people stories than for molecule stories. But if and when I get to write about people, there has to be some science in it, such as the chemistry that attracts people to each other, or the genetics that makes them the individuals that they are.

The link between chemistry and genetics is of course sex; the former leads to it and the latter results from it. So this is my ramshackle excuse for collecting various stories ranging from the chemistry of attraction through to genetics under the heading of sexy science. Be warned, however, that there are some stories that I just found sexy for no particular reason.

Feel the Heat

Let's be very unsubtle and start the sexy part of the proceedings with steaming hot sex. Voyeurs will need a microscope, though, as we are talking microbial sex.

"Plaisir d'amour ne dure qu'un moment,
chagrin d'amour dure toute la vie."

The joy of love only lasts for a moment, while the sorrow of it may last a lifetime, as the French song by Jean-Pierre Claris de Florian wants us to believe. Even if the *chagrin d'amour* may last for ever, there is still a major biological advantage to be gained from using sexual reproduction. In comparison with simple cell division accompanied by the odd random mutation, the mixing of the parents' genes allows for a greater genetic variation combined with better stability of the collective gene pool of the population. In a sense, this is easier for us higher organisms, as our genetic material is organized in many pairs of chromosomes, so we just have to make sure we get one from each parent. Which of the two copies we receive, that is the lottery of life. The sperm that wins the race against millions of others may carry Granny's musicality or Granddad's big ears, and these genetic dispositions may override the corresponding genes from the maternal side or may lose out and remain silent for a generation: that's the name of the game. For bacteria, such a gene lottery is far more difficult to organize, as their genome normally comes as one double strand of DNA. Gene transfer only works if short DNA fragments can be cut out of the genome and then relocated with the help of viruses or mobile DNA rings called plasmids. Hence, for many microbial species, in-

The Birds, the Bees and the Platypuses. Michael Gross
Copyright © 2008 WILEY-VCH Verlag GmbH & Co. KGaA, Weinheim
ISBN 978-3-527-32287-9

cluding the extremely heat-loving (hyperthermophilic) archaebacteria, sex used to be taboo.

In 1996, however, Dennis Grogan of the University of Cincinatti found the right kind of temptation to bring the heat- and acid-loving microbes of the species *Sulfolobus acidocaldarius* into intimate contact. He grew various mutants of the archaebacterium, which – apart from the basic nutrients provided in the minimal medium – each needed one amino acid or coenzyme as a food additive to thrive. For instance, a mutant strain that lost the ability to produce the amino acid histidine from scratch would need a histidine supply in the growth medium. Then he mixed two of these mutant strains and incubated them on a medium containing neither of the required additives – it mainly consisted of the amino acid glutamic acid and dilute sulfuric acid. Only through a gene transfer from one strain to the other could a new strain arise which would be able to thrive on a medium with no additives. In fact, Grogan found such "cured" colonies quite frequently, while a control experiment where the strains were not mixed but incubated separately on minimal medium only very rarely led to reverse mutations re-establishing the additive-independent wild type. Even more surprisingly, the gene transfer, or microbial sex, was obviously happening at temperatures of up to 84 °C.

This was the first time that an exchange of genetic material between bacterial cells has been demonstrated at such high temperatures, and also a first for thermophilic archaebacteria, and it opens up interesting opportunities and poses new questions. Grogan believes that *Sulfolobus* needs the transfer as a repair mechanism in case its DNA suffers heat-induced damage. In a similar way, primitive microbes in the early days of life could have coped with genetic defects caused by UV irradiation, which was then much stronger than it is now. Since those times, archaebacteria are believed to have changed less than any other group of organisms; thus their habits of living and loving may well be a window into the past.

(1996)

Further Reading

M. Gross, *Life on the Edge*, Plenum, 1998.

What Happened Next

Microbial sex has continued to worry us in the last few years. Not so much the steamy kind happening among hyperthermophiles, but the one happening in hospitals where antibiotics are abundant and microbes swap resistance genes in order to survive, which leads to multidrug-resistant strains, including the hospital superbug MRSA.

Mum's and Dad's

> "I rarely get the chance to use poetry in my articles. The Goethe lines that fitted this one so nicely were often quoted in my family when I grew up, so I jumped at the opportunity to peg a science story to them."

Vom Vater hab ich die Statur,
des Lebens ernstes Führen,
vom Mütterchen die Frohnatur
und Lust zu fabulieren.

In this way, Johann Wolfgang von Goethe analyzed the redistribution of genetic traits at his own conception. He believed he had inherited his body shape and seriousness from his dad, while his sunny side and desire to tell stories must have come from his mother.

He isn't the only one to be interested in such questions and, as we now know quite a lot about the rules of the genetic game, there is a better chance to track down just where those big ears came from, even though there are still some devilish details to be investigated.

Mum and Dad each contribute 23 chromosomes to the set of 46 that we have, one of each pair, including the potentially lopsided pair of sex chromosomes. This arrangement is fully deterministic. Any deviation from this rule will result in serious disorders, such as Down's syndrome. A major part of the lottery of life, however, lies in the question of which of the two copies of each gene and of each chromosome Mum and Dad will pass on: The one from Granddad or the one from Granny. As each egg and each sperm only carries a single set of 23 chromosomes, hard choices have to be made.

The Birds, the Bees and the Platypuses. Michael Gross
Copyright © 2008 WILEY-VCH Verlag GmbH & Co. KGaA, Weinheim
ISBN 978-3-527-32287-9

In theory, entire chromosomes might be picked from one side or the other, but there is an additional mixing process known as crossing over, which enables Granddad's and Grandma's chromosomes to hook up and swap large stretches of their genetic material, such that in the end there is no telling whose genes will end up in Dad's sperm or in Mum's egg. This is why siblings (apart from monozygotic twins) are all different.

But after the balls of that particular lottery have rolled out, and the Winning Sperm has fertilized the Egg of the Month, who is to decide which version of a given genetic trait will prevail, Mum's or Dad's? In certain cases studied extensively by classical genetics, the answer lies in the nature of the gene. Dominant genes will prevail even if only inherited from one parent, while recessive traits require matching copies from both parents in order to manifest themselves in the offspring.

In the 1990s, however, researchers found that there is more to the lottery of life than just crossing over and the tug-of-war between dominant and recessive genes. In some cases, they found, identical genes will have different effects depending on whether they came from Mum or from Dad.

All this goes back to a surprising discovery which Azim Surani made at the Institute for Animal Physiology at Cambridge back in 1984. He wanted to know why mouse embryos created from unfertilized eggs didn't survive to the end of the gestation period. Therefore, he started manipulating freshly fertilized eggs, where the nuclei originating from the egg and the sperm are still separate and referred to as pronuclei. He replaced one of the two pronuclei – containing the single chromosome set of one parent – with the corresponding structure from a different fertilized egg. He found that the manipulated eggs only developed normally if they continued to possess one pronucleus derived from a male and one from a female. Those that ended up with two chromosome sets from a male or two from a female perished after just a few cell divisions.

Thus, it was obviously not just the number of chromosomes that was important for development to work, nor the combination of sex chromosomes, as the combination of two female pre-nuclei would have resulted in a normal combination of two X chromosomes, as in every female conceived the normal way. In addition to these numerical issues, the presence of some unidentified contribution from each parent, possibly a kind of label indicating the provenance of genes or chromosomes, appeared to be crucial for successful development. Such a labeling process would have to be reversible, as any maternal gene could end up in the son's sperms, and conversely, a paternal gene could become a maternal one in the daughter's eggs.

Surani suspected that embryonal development used only the maternal or the paternal version of certain genes, while the other one is labeled "Do not use." This was why the embryos in his experiment were doomed if they had only maternal or only paternal chromosomes. Sure enough, researchers soon found detailed evidence that at least 15 genes in mice and in humans carry such labels. The phenomenon was dubbed "imprinting."

To work out details of how imprinting works, researchers had to find more subtle and targeted methods. Barry Keverne and his coworkers at Cambridge created patchwork embryos in which only a few cells had only one parent, while the others had the normal set of chromosomes from both parents. Provided that at least half of the cells in an early phase embryo were normal, development could proceed. Intriguingly, mice with an overdose of maternal chromosomes turned out to have large heads and small bodies, while those with excess of paternal chromosomes showed the opposite effect.

In terms of sociobiology, one may be tempted to interpret these tendencies as a conflict of interest between the parents. The genes from the father would thus work towards creating a large baby without worrying about how much energy that will cost the mother. Maternal genes however would be more economical with the mother's energy and give the child a bigger brain to compensate for smaller stature. Unfortunately, however, detailed analyses of mutation rates (which should be high if there was a conflict between parents) failed to confirm this hypothesis.

The conspicuous differences in head and brain sizes inspired Surani and Keverne to check precisely in which parts of the brain the cells with purely paternal or purely maternal chromosomes end up. To this end, they labeled the manipulated embryonic cells in such a way that they would later be able to locate their descendants in the brain of the grown-up mouse.

The surprising result: No matter how many or few cells carried maternal genes only, their descendants dominated the same brain regions, namely those that are in charge of higher functions such as learning and memory. By contrast, the cells are excluded from certain other areas related to instincts such as feed and sex drive. And yes, you guessed it, the purely paternal cells go exactly the opposite way. "No learning and memorizing for me, thanks. I'll stick with the booze and the girls," Dad's imprinted genes seem to be saying.

Let's remember, though, that we are still looking at mice here. Would these insights be transferable to men, could they confirm that Goethe got his enjoyment of storytelling from his mother? In fact, Keverne must have been aiming at such questions, because his main research field is cognitive skills of primates. He refers to those brain areas that appear to be linked to maternal development genes and fulfill complex cognitive tasks as the "executive brain," pointing out that this is the part that has grown the most on the evolutionary path to modern humans. He contrasts this with the "emotional brain" connected to paternal genes, which is in charge of hormone-driven, instinctive actions, and which has shrunk somewhat during our evolution.

If we use this division of the brain derived from developmental genetics to look at the social behavior of different primate species, there are some intriguing connections to be made. For instance, the larger the groups that primates tend to form, the smaller their emotional brain. Thus, too much instinct seems to be bad for politics. Conversely, the size of the executive brain increases with group size, appearing to take control out of the hands of the receding emotional brain.

Similarly, our love life seems to have been partially transferred from the realm of the hormones to the realm of the more conscious senses. The size of the executive brain increases with the length of the mating season. Unlike most mammals, we can engage in sex all year round, but we can also decide not to.

Some press reports were quick to conclude that fathers get short-changed by genomic imprinting. However, it remains unclear whether imprinting really has a role in handing on specific intellectual abilities from one generation to the next. Typical human cognitive skills such as language, planning, and creativity depend on a complex network of genes, only some of which will carry imprinting signals.

Moreover, investigations on individuals with chromosomal aberrations, such as Turner syndrome (where a single X chromosome is the only sex chromosome present, and may be inherited from father or mother) have cast doubt on such immediate conclusions, as Turner patients with a paternal X chromosome tend to do better in social and cognitive skills than those with a maternal one. Thus, much more detailed and subtle (not to mention less destructive!) ways of analysis will have to be found before scientists can really work out where those storytelling skills came from.

(1997)

Further Reading

W. Reik and J. Walter, *Nat. Rev. Genet.*, 2001, **2**, 21.

What Happened Next

As of 2007, it is not clear which genes in the human genome are imprinted. A study published in March 2006 provided a shortlist of 600 candidate genes, which researchers will have to test individually.

Gold Clusters Shine Brightly

Throughout history, and even today, sex appeal can be enhanced with a bit of bling bling. Shiny things like gold and diamonds are often considered to be lazy bones when it comes to chemistry, but this may just be a question of looking a bit closer. After all, silver and platinum can be very useful in photography and catalytic converters, respectively. So let's see whether gold can add a bit of bling to chemistry.

Although gold is known to be the noblest of all metals, it can in fact form chemical compounds. For instance, if chemists heat finely powdered gold with chlorine gas, they obtain a salt (gold chloride), which they can dissolve in water, crystallize, and convert to numerous other gold compounds with different oxidation levels. Thus, it might almost be mistaken for a normal chemical element, had there not been the historical equivalence of gold with wealth and power, leading to the obsessive attempts of alchemists to "make" gold. An obsession which on one hand drove the development that eventually led to the science of chemistry but on the other hand burdened it with a dubious heritage.

Nowadays, as gold does not build or destroy empires any more, and its power has faded next to the importance of other resources such as oil and uranium, chemists can return to the interesting chemistry of this metal without being suspected of practicing alchemy. Nevertheless, there are not that many researchers who study gold compounds. But what they find is often surprising and intriguing.

Copper, silver, and gold are meant to be similar, as they live in the same column of the periodic table. Gold, however, being the heaviest member of the group and having a very complex electronic structure, remains a bit of an outsider. One incarnation that has proven particularly interesting is the monovalent gold ion (Au^+ to the chemist), which is analogous to silver in simple compounds, such as silver nitrate or silver bromide, but behaves very differently. For instance, it is not stable in solutions when it is on its own, but it can be stabilized with small organic molecules (ligands) that can shield it. The resulting gold compounds often contain several Au^+ ions held together by weak bonds.

Quite a few of these compounds, including the phosphane–gold complexes, can be made to fluoresce when excited by UV light. (That is, they send out light of a longer wavelength than the light they receive, which enables them to "convert" invisible UV light to visible light. Fabric brighteners work on the same principle.) Intensive investigation of these phenomena during the 1980s has shown that most gold compounds can only shine in the crystalline form. In rare cases, which were only discovered in the 1990s, solutions of the compound may be luminescent, too. And then there is a third way, which was only reported in 1997. When Ella Fung, working with Allen Balch at the University of California at Davis, tried using chloroform to wash crystals of a gold compound she had prepared, they began to shine, although neither the dry crystals nor a solution of the substance was luminescent.

Although those crystals had not been deliberately irradiated with UV light, the researchers soon came up with the hypothesis that an inadvertent exposure, e.g. from the room lights, could have stimulated this effect. Systematic investigation using a UV lamp confirmed that UV light can bring the crystals into an excited state, which is stable for hours and only starts releasing the energy in the form of yellow light when the crystals are at least partially dissolved. Although the luminescence fades quite rapidly, it can be reactivated by a renewed irradiation with UV light. And if one evaporates the solvent used for washing the crystals and irradiates the solid, this will luminesce upon contact with chloroform as well. Different solvents can trigger this effect, but its intensity increases with the solubility of the gold compound in the solvent. These findings, along with the observation that neither chemical reactions nor mechanical stress can make the compound shine, lead to the conclusion that the process of dissolving the crystalline material is crucial.

It remains unknown, however, what the excited state looks like and how the luminescence phenomenon arises. Structural investigations carried out by Balch's group revealed a stacking of the ring-shaped gold complexes, with the triangle of gold atoms always in the same orientation. As this crystal lattice, with its unusually short distances between the metal centers, is as unique as the solvent-induced luminescence, researchers are wondering whether the former might be the cause of the latter. The lengthy columns with their core made up of nanoscale "Toblerone™ bars" of pure gold might be the key for understanding the phenomenon. According to one hypothesis, the UV light might induce a charge separation resulting in electrons migrating along the Toblerone bar, which may act like a nanoscopic gold wire. Electrons may then get stuck at irregularities in the crystal, where they would store the energy corresponding to the charge separation until they are released by the dissolution of the crystal. The reunification of electrons and positively charged metal centers would result in the luminescence phenomenon.

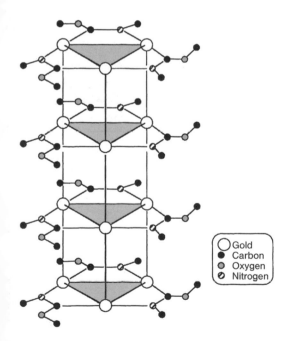

◯	Gold
●	Carbon
◉	Oxygen
◐	Nitrogen

Figure 11 Gold clusters adopt the shape of a triangular prism, like a Toblerone ™ bar.

Thus far, this is only speculation, and the same holds for the potential applications that have been discussed in the literature, including sensors for solvent vapors, energy storage, and photochemical switches. These dreams will only materialize when the phenomenon can be understood and directed into technically defined pathways. And, of course, only if there are no other materials that can fulfill the same function at less expense. Nevertheless, these findings have thrown up intriguing questions, and they have demonstrated that noble elements don't have to be boring elements.

(1997)

Further Reading

M. Gross, *Travels to the Nanoworld*, Perseus, 1999.

What Happened Next

I have no news from the shiny gold clusters, but there have been some other intriguing developments in the chemistry of this noble element. In 2006, Stephen Hashmi at the University of Heidelberg reported gold catalysts, and in 2007 F.D. Toste and colleagues described a clever way in which gold catalysts can even be made to distinguish between left- and right-handed versions of their substrates. This kind of discrimination, known as chiral catalysis, is one of the key concerns of modern catalysis research, as chirality is important in the cell and thus also in drug development.

The Green Spark

Lights can be sexy, obviously, otherwise there would be no red light districts or candle-light dinners. To molecular biologists, however, a molecular candle that glows bright green is one of the sexiest discoveries of the last couple of decades, which has also led to some cool new technologies (see page 193).

Bioluminescence – converting metabolic energy into light – is a fascinating phenomenon, and more common than you may have thought. Intriguingly, evolution re-invented bioluminescence more than 30 times independently. Apart from the well-known fireflies and glow-worms, a wide variety of fish, jellyfish, bacteria, and even mushrooms also glow in the dark. Although much of the fundamental work on bioluminescence was based on the light organs of insects, the greenish glow you are most likely to see around a biology lab these days comes from the molecular systems of the jellyfish *Aequorea victoria*.

Why is it that the luminescence system of these wobbly creatures has become so useful? This has to do with the fact that the jellyfish has two separate steps leading from the perception of a chemical stimulus to the production of green light. In the first step, a protein called aequorin responds to a signal (carried by calcium ions) by sending out light. As one can verify by exposing the purified protein to calcium ions in the test tube, this light is not green, but blue. In the jellyfish, however, you will never see the blue light, as it is immediately absorbed by a second protein, which converts the energy into green light. This latter molecule has become universally known as the green fluorescent protein, or GFP. In order to produce its characteristic green glow, it needs no other molecules. It is activated by blue or ul-

The Birds, the Bees and the Platypuses. Michael Gross
Copyright © 2008 WILEY-VCH Verlag GmbH & Co. KGaA, Weinheim
ISBN 978-3-527-32287-9

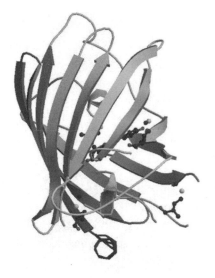

Figure 12 Structure of the green fluorescent protein (GFP), as revealed by X-ray crystallography. The arrows represent beta-pleated sheets, which make up the barrel-shaped exterior of the protein. The atoms drawn as balls in the inside of the barrel belong to the structure responsible for the light reaction, the chromophore.
(From the RSCB Protein Data Bank)

traviolet light alone, even if it has been produced by genetic engineering and has never been anywhere near a jellyfish.

This remarkable property of GFP leads directly to an important application, which was first suggested by Martin Chalfie and coworkers from Columbia University in 1994 and went on to become general laboratory practice in a matter of months. Researchers often want to transfer a gene from one organism into another, so that the latter starts producing the protein encoded by the gene, but it would useful if there was a simple way of knowing that the transfer had occurred. By simply coupling the gene for GFP with the gene of interest, attempting the gene transfer, and placing the transformed cells underneath a UV lamp (which is available in every molecular biology lab), they can know immediately. If the cells produce green light, the transfer was successful. It's as simple as that. There have been other luminescence assays before, but they all required additional chemicals, which would have to be transported into the cell through the membrane. If the cells refused to take up the required substrate for whatever reason, this would have led to a false negative result.

Thus, GFP was recognized as a gene marker of unique potential, and within a year of the first paper on its application appearing, its gene was commercially available as a molecular biology kit. Biophysicists, too, enjoyed playing around with this protein. They tried to make it even more useful by making it shine in different colors, and they can even make it blink rhythmically, as we will see below.

Basic research into the structure and mechanism of GFP had trouble keeping up with the pace of progress in molecular biology. Only in 1996 could the molecular structure of the protein be revealed, and it was found to involve a novel and unusual protein architecture. The outer coat of the protein is a completely symmetrical and remarkably large barrel made up from 11 strands of a beta-pleated sheet. The ends of the barrel are covered with lids made of short helical segments. Within the cavity of the barrel, there is a much longer helix, whose axis coincides with that of the barrel. And bang in the middle of that helix, like the filament in the middle of a light bulb, we find the luminescent part of the molecule – the chromophore. It is formed by an unusual chemical reaction taking place between neighboring amino acid residues within the protein chain, as the protein folds for the first time.

Knowing the three-dimensional structure of this protein is particularly important, as this structure is thought to catalyze the very unusual process by which the chromophore is formed. As the formation of the chromophore has worked nicely in every host organism that has ever been tried, researchers suspected that it was an autocatalytic reaction, that is, a reaction catalyzed by the same molecule rather than by some other, as yet unknown, enzyme.

However, the direct proof of this hypothesis was only accomplished in 1997 by Brian Reid and Greg Flynn at the University of Oregon. They produced the protein in the form of not-yet-folded polypeptide chains clumped up to so-called inclusion bodies in *Escherichia coli*. After purifying and dissolving the inclusion bodies, they could observe GFP molecules which had never been folded, let alone active, fold up and carry out the chemical modifications that lead to formation of the chromophore – a process lasting several hours.

The known structure also allows us to understand or predict how mutations, i.e. the exchange of individual amino acids in the sequence, will affect the luminescence of the protein. Although the green light that is produced is essentially monochromatic (i.e. all photons have roughly the same wavelength), the energy intake can occur in two different ranges of the spectrum. The natural protein from jellyfish best absorbs light that is violet to near ultraviolet (around 396 nanometers), but it can also use bluish-green light with wavelengths around 476 nanometers. Researchers have worked out that the former involves the electrically neutral state of the chromophore, while the latter requires a negative charge in this part of the molecule. By introducing subtle changes in those regions of the protein known to be close to the chromophore, they can now influence its charge state and thus produce GFP variants with tailor-made optical properties. For instance, practical reasons would call for a stronger absorbance at 480 nanometers, because there are lasers with this specific wavelength that could be used to illuminate protein molecules with very high precision in both space and time.

Two of these variants also led to the discovery of blinking in GFP, made in 1997 by the group led by W. E. Moerner at the University of California in San Diego. In order to observe individual molecules of GFP separately, the scientists had brought the protein molecules into a porous jelly, so that they could not move around but could still be accessed by light and chemicals. They irradiated the trapped molecules with light of the appropriate wavelength, and filmed the luminescence through a microscope. Provided that the molecules are separated by a distance longer than the resolution of the microscope, the light emission of individual molecules can be recorded separately.

This way, Moerner and his coworkers could establish that each molecule fluoresces for a few seconds, then goes dark for a few seconds, and then starts to shine again. This "blinking" lasted for a couple of minutes, and ended when the molecules had sent out around one million photons, at which time they "retired" and remained permanently inactive. However, even this "retired" state could be reactivated by irradiation with light of a shorter wavelength (higher energy) than that used in the original experiment.

Obviously, the protein can occur in three different states: the active one, the permanent dark state, and the intermittent dark state. While physicists are confident they can explain the first two states, it is the dark state between light pulses that made them scratch their heads. Current theories are still rather hypothetical, and I won't go into any further details here.

Apart from the novelty and intellectual challenge, this experiment also has practical importance. It is probably the first molecular optical system that can be switched at room temperature between two different states which can be read out from individual molecules. Thus, a molecule which responds to a short illumination with blinking, would code for a binary 1, while one that is in the dark state would code for 0. While similar switching can be performed on the protein bacteriorhodopsin, in that case reading out the information stored in the molecules would require a measurement of absorbance. The advantage of GFP being luminescent and detectable on a single molecule level suggests possible applications in many areas of future opto-electronic devices, through to the elusive protein-based computer.

Closer to the present-day reality, however, application of GFP and its variants has become common practice in many laboratories. And it is not only unicellular microbes that glow in the dark to reveal the presence of a transferred gene. In some places, whole plants and animals will light up, in others just certain parts of them, such as tumors, for instance. Karl J. Oparka and his coworkers at the Scottish Crop Research Institute in Dundee have coupled plant viruses with the GFP marker in order to study how the virus spreads through infected plants, and which factors may influence its progress. Plant physiologists in Japan have used GFP to study the heat shock reaction in rice plants.

GFP shines in animals as brightly as in plants. Pioneering studies involved GFP expression in fruit flies (1994) and in zebra fish (1995). Using a GFP variant with increased luminescence intensity, Masaru Okabe and his coworkers at the University of Osaka, Japan, could for the first time produce mice that, under the UV lamp, shone up bright green through and through. The researchers had smuggled the gene into fertilized eggs, and it had indeed been passed on to all the mature cells of the body. Moreover, it was passed on to a new generation of green fluorescent mice.

Such studies, of course, are not meant to address the problems you may face when you have misplaced your lab mouse in the dark. Mainly, they promise to deliver new insight into the development of mammalian embryos, along with new diagnostic tools for medical applications. GFP could be used to track the spread of metastases. More generally, whenever cells of different origin are observed together, fluorescence labeling would provide a convenient way of distinguishing them. And gene therapy methods of the future will certainly benefit from the option of controlling and targeting the therapeutic gene using a GFP marker. In a pioneering experiment, GFP could be introduced into and kept stable in a cell culture derived from human melanoma cells.

More and more complex tools will be built with and around GFP. In one recent example, Atsushi Miyawaki and his coworkers at the University of California in San Diego have combined two differently colored variants of the protein in one construct. They surround a calcium binding unit made of the protein calmodulin and the peptide M13. As soon as the calmodulin binds a calcium ion, the geometry of the construct changes in such a way that the variant which absorbs and fluoresces at shorter wavelengths can pass on the energy to the other GFP moiety, which will then emit light of a characteristic wavelength. Depending on the binding efficiency of the calmodulin variant used, this method can measure calcium concentrations ranging from a hundredth down to 10 billionths of a mole per liter, with a spatial resolution that allows researchers to compare the concentrations between different compartments of a single cell.

Biophysics, medicine, developmental biology, molecular biology ... there seems hardly any area in the wide field of the life sciences that the green spark has not yet lit up. GFP serves as a model substrate for molecular chaperones, as a probe for measuring biophysical parameters within cells, and as a luminescent trace to follow transport pathways in cells or organisms. While other molecules have had similarly explosive careers, none has spanned so many different aspects and found practical applications in so many different areas. The green light seems to say "go" in more than one sense.

(1995)

Further Reading

M. Gross, *Light and Life*, Oxford University Press, 2003.

What Happened Next

As of 2007, GFP and various products derived from it shine in many colors in many places around the world. A quick Google search turned up 1.74 million hits for it. There are a few very strong candidates for future Nobel prizes in this story.

Read My Lips

I get all kinds of feedback on the pieces I publish, but only once in my life did I receive an invitation to the premiere of a theatre production inspired by one of them. And then I couldn't make it, so I'll never know how my work translates to the stage. The text in question was an opinion piece about the McGurk effect and dubbed movies, and it also triggered a lively correspondence when it first appeared in *New Scientist*. Here's a director's cut version:

This is absolutely no joke. Without normally being aware of it, we practice lip-reading in normal conversation. This was proven by Harry McGurk and John MacDonald at the University of Surrey back in 1976. Confronting listeners of different age groups either with a recorded voice only or with the same voice together with a video showing non-matching lip movements, and asking the subjects to repeat what was said, they found that up to 92% of the answers were wrong when the conflicting visual input was present, while nearly all of them were correct when it wasn't. For instance, subjects hearing "ba-ba" but seeing a lip movement that corresponds to "ga-ga," reported they heard "da-da," indicating that the perceived speech is a fused signal of the auditory and the visual input.

The McGurk effect, as this phenomenon came to be called, has been quoted more than 80 times in scientific journals ranging from *Advanced Robotics* to *Science*. Follow-up studies have addressed all kinds of specific cases, including audiovisual cross-dressing ("... female faces and male voices in the McGurk effect"), intercultural comparisons ("Cultural and linguistic factors in audiovisual speech processing: The McGurk effect in Chinese subjects"), and speakers

The Birds, the Bees and the Platypuses. Michael Gross
Copyright © 2008 WILEY-VCH Verlag GmbH & Co. KGaA, Weinheim
ISBN 978-3-527-32287-9

standing on their heads ("Perceiving speech from inverted faces" – my favorite title in the list).

However, a large-scale experiment on this effect has been going on in many countries virtually unnoticed – to my knowledge no-one has even bothered to collect the results. Exposure of millions of individuals to McGurk style experimental conditions is going on in all those countries where foreign films are routinely shown in dubbed versions, as is the practice in Germany and France for instance, but neither in the Netherlands nor in Britain. Amazingly, most German spectators do not find it irritating that the lip movement does not match the sound, suggesting that the McGurk effect can be switched off, if a year-long habit tells the brain that the lip-signals in such cases are nonsensical. In contrast, when found in a situation where the dubbing goes against the habit, e.g. when seeing a German actor dubbed into French, German viewers would strongly object to the "wrong voice."

From my own experience of moving from a dubbing to a non-dubbing country (Germany to Britain), I know that the unconscious lip-reading inactivated by habitual viewing of dubbed films comes back after only a few months in a non-dubbing country, even with only a moderate exposure to cinema and TV. Nowadays, seeing dubbed material is quite enough to drive me up the wall, no matter from which language into which other, provided I know at least one of the two. For instance, watching a Swedish film (with actors I did not know, so I had no preconception as to what language or what voice they should have) dubbed into German was absolutely unbearable, exclusively because of the outrageous mismatch of lip movements. (I do not understand a single word of Swedish, so I didn't mourn the loss of the original language as such in this case.) Other traumatic experiences include the German version of the French film "Cyrano de Bergerac." The story is – as you may remember – about a guy with a very long nose winning a woman's heart by the combined use of beautiful language and a man of straw shorter in both nasal protrusion and linguistic ability. The dialogue, which is indeed stunning in the original, turned into unbearably bad verse in the German version, thus making the whole story appear ridiculous.

Even with languages of which I have very little knowledge, such as Italian or Dutch, I feel that something comes across through the spoken language even while I am reading the subtitles – the Dutch movie

"Antonia's line" was a perfect example for this. I found the original language so poetic I stopped reading the subtitles, relying only on the similarity with German to catch the contents of the dialogue. Similarly, I enjoyed learning some basic Italian from "Il Postino," although it was irritating to see Philippe Noiret dubbed from French into Italian. (Some of the recent films mixed from half a dozen different countries like cheap EU wine, don't actually seem to have such a thing as an undubbed original version.)

Considering, however, that most of the European dubbing business goes on from English into German (and French, Spanish, etc.), while every German moviegoer knows at least as much English as some of Hollywood's highest earning action movie protagonists, and certainly more than I know Dutch, I don't see any point in this whole business. Consumers certainly lose part of the natural audiovisual experience, good movies can be disfigured beyond recognition, and even bad films cannot win. (German TV consumers tend to argue at this point that Starsky and Hutch are funnier in German than in the original version, which I haven't checked. Even if this is true, it is still the exception which proves the rule.)

You may at this point wonder what I am talking about – if you only know one side of this cultural divide. If you live in a non-dubbing country and have never watched a dubbed film, try getting a Swedish children's video like, for instance, "Pippi Longstockings." These should be dubbed in most countries, and the close-ups of children speaking rather loud and with very expressive faces provide a good introduction into the phenomenon. In contrast, if you live in a dubbing country and have never realized that the wrong voices are coming out of your TV set, spend your next holiday in some country where films are dubbed into a different language (France or Spain would do nicely) and watch some ordinary American films with the "wrong" language. Then consider that the voices are exactly as unmatched as the ones emerging from your TV set at home.

In either direction, you will be surprised – I certainly was. On the scientific side, of course, this mega-McGurk experiment carried out with millions of unaware human guinea pigs provides a unique chance. Experimental psychologists should investigate how viewers manage to switch off the lip-reading without even being aware of what they are missing. For cultural reasons, however, I should like to call for a ban on dubbing, or – seeing the lack of a suitable authority to enforce this – for a boycott of dubbed versions.

(1997)

Further Reading

Just google "McGurk" to find lots of weird and wonderful insights on lip-reading.

What Happened Next

Even though this was one of my more successful pieces in terms of impact and feedback, I haven't followed this field any further, so I don't really know. But if you do research in this field and come up with something exciting, do give me a shout, and I will read your lips.

You Taste So Sweet

Sex can make life sweeter, and sweets can make it sexier, so there is clearly a meaningful connection here that could serve as an excuse to include the sweet little story about proteins that taste sweet. On top of that, there are also proteins that alter our perception and make other things taste sweet. Which again is a lot like love making us blind. So without further sweet talking, here come some taste-boggling sensual experiences.

People in West Africa have used the serendipity berry (*Dioscoreophyllum cumminsii* Diels) as sweetener for centuries. Not only is the taste of slimy pulp very intense, but it also tends to be very long-lasting.

Scientists of the Western world only became aware of these berries in the 1960s, when concerns over potential cancer-promoting effects of cyclamate triggered a hectic search for replacements. (By now, cyclamate has been cleared of this suspicion, but has lost most of its former market share to aspartame).

In 1971, researchers at the Monell Chemical Senses Center at Philadelphia isolated the active ingredient from the tropical berries. Surprisingly, it turned out to be a small protein, with a molecular weight of 10,700. In honor of the institute where it was discovered, the researchers christened it monellin. Compared with an equal weight of household sugar, monellin is 3000 times sweeter. For the most effective synthetic sweeteners, aspartame and saccharine, this figure is 200 and 450, respectively.

It is now known that there is a small but quite exquisite family of proteins that display a very strong sweet taste. For a long time, they were regarded as nothing but a curiosity, but in the 1990s people became aware of their potential. Thaumatin was the first such protein to be approved as a food additive in North America and in Europe.

But how do these proteins achieve this miraculously strong sweetening effect? In the mid-1990s, researchers hoped to answer this question by solving the molecular structures of three such proteins in detail, namely thaumatin, brazzein, and monellin. Their hopes were crushed when they found that the three proteins had no structural resemblance whatsoever. Confusingly, each of them is more similar to a different, taste-inactive protein than to the other two. Brazzein, for instance, appears to be related to a family of proteins that have a broad range of functions in plants, from defense through to metabolic enzymes.

Mutation studies on brazzein have come up with two sites linked to the taste activity. Frustratingly, the two sites are on opposite ends of the molecule, so they cannot possibly both interact at the same time with a taste receptor on the tongue.

How, exactly, the sweet proteins bind to the corresponding sensors remains unclear. Part of the blame can be assigned to the fact that the sense of taste is very poorly understood in molecular terms (see the chapter on the bitter taste receptor, page 114).

One thing seems certain, though. The surprisingly intense sweetness of these proteins appears to reflect a very tight binding to their target molecule. This is also confirmed by the observation that the sensation of sweet taste persists for minutes or even hours. Ordinary sugar is a rather poor signaling molecule by comparison. The smallest quantity of sugar that we can taste is by orders of magnitude larger than the minimum quantity of hormones necessary to achieve an effect.

Thus, it appears possible that sugar doesn't in fact bind very specifically to a receptor. Instead it might have a more general effect on a signaling element further down the taste perception line, such as an ion channel.

The characteristic strong binding and long-lasting effect are also found in a second group of "tasty" proteins, namely the taste-modifying proteins. Proteins such as curculin and miraculin have no taste of their own, but ingestion of a minute amount of such proteins will make sour foods taste sweet for hours.

All these "miraculous" substances are potentially interesting for the food industry. While the complicated extraction procedure from tropical fruit makes them quite expensive, this is compensated by the extremely high effectiveness at very low doses. For the same reason, the sweeteners could also be attractive as dieting aids. The amount of protein equaling a spoonful of sugar is so small that its calorific value is negligibly small.

What Happened Next

Amazingly, there haven't been any major new revelations in this field since I wrote about it in 1998. There are still only five sweet proteins and two taste-modifying proteins known, and their mode of action has remained elusive.

Stop press: In September 2007, researchers at Nagoya City University in Japan reported mutational studies of curculin, combined with tasting analysis. The results seem to suggest that different parts of the molecule interact in different ways with the sweetness receptor, thus producing the separate sweetening and taste-modifying effects.

A Matter of Taste

Part of the problem with understanding the sweet proteins discussed above lies in the fact that the sensory perception of taste and smell is a topic that modern science knows embarrassingly little about. Only since the late 1990s have researchers started to get a grip on this field, creating some exciting stories of discovery along the way, which I have covered on various occasions.

Have you ever asked yourself, while enjoying a delicious meal: "I wonder how taste works"? OK, probably not, as people tend to be either science nerds or foodies, but rarely both. And perhaps it's better if you haven't, because the answer would have to be that even the scientists who study it don't really know. Of all our five senses, taste is arguably the one that's least understood. The masses of genetic information flooding in with the human genome project, however, have now helped scientists at least to start closing this embarrassing gap in their knowledge.

There are five basic tastes that the taste buds in our tongues and palates can distinguish: salty, sour, sweet, bitter, and umami (the taste of glutamate, which is contained in meat and soy sauce). It is symptomatic of the state of our knowledge that the latter was only discovered in 1999. The perceptions of salty and sour are relatively easy to explain, as each only involves single type of charged atomic particles (called ions), which most cells easily recognize. Umami is about one kind of molecule only, sodium glutamate. But sweet and bitter have remained elusive, because there are many different substances that trigger these taste sensations without necessarily resembling each other. This is particularly striking for bitter substances. They can have

The Birds, the Bees and the Platypuses. Michael Gross
Copyright © 2008 WILEY-VCH Verlag GmbH & Co. KGaA, Weinheim
ISBN 978-3-527-32287-9

many different structures, which to a chemist have nothing whatsoever in common, but still all taste the same.

To a cell biologist this implies that on the surface of a cell that specializes in bitter taste, there must be many different probes, called receptors – one for each kind of bitter tasting molecule recognized. All these receptors should talk to one signaling station that forwards the soundbite "something bitter" to the nerve cells and eventually to the brain. By analogy to other sensory perceptions and to signaling events found in hormone response, scientists expect the signaling station to be a protein from a certain large family, the G proteins. One G protein putatively involved in taste signaling was identified in 1992 and named gustducin (derived from the name of the G protein involved in vision, transducin). But until 1999, not a single receptor for taste had been identified. And when, eventually, a pair of receptors was found in 1999, they turned up only in cells lacking gustducin. It was a situation a bit like digging a tunnel from both ends but failing to meet in the middle: Rather than building up a complete signal transduction pathway, researchers now had receptors without the corresponding G protein, and a G protein lacking a receptor.

Things only started to get better when the team of Charles Zuker at the University of California in San Diego began using genome data in the search for taste receptors. They knew that the ability to taste a certain bitter substance called PROP is genetically determined by an unknown gene in a certain region of chromosome 5. When the sequences of this area became available in 1999, they used computer-assisted searches to check whether it had any genes coding for receptors that look like those that typically communicate with G proteins. They found one, which they called T2R1, and then went on searching for genes similar to this one. The search resulted in 19 further candidate taste receptor genes, which implies that this family could have up to 80 members throughout the genome. Unlike the receptors found in 1999, these are plausible partners for gustducin, as the proteins are produced in the same cells.

Indeed, Zuker and his coworkers have shown that the members of the T2R family are not just candidates but that they do represent fully functional receptors, recognizing one bitter substance each. This could be established in cell cultures derived from mouse taste cells, because the corresponding area of the mouse genome is very similar to the human one, and the mouse receptors could be readily identified and matched to the human analogues.

Groundbreaking research findings in a particular field are like London buses, only worse: You wait for years or even decades, and then you have five in a row. Thus, independently of Zuker's group, a team at Harvard University also reported having found a family of bitter taste receptors as a result of screening through the available genome sequences of mouse and human DNA. The group of John Carlson at Yale found molecules which are probably taste receptors in the fruit fly *Drosophila melanogaster*, and another paper reported a receptor for umami.

Obviously, the time has come for a better understanding of taste perception. And it was about time, really. Apart from the downright embarrassing lack of scientific explanation for such a fundamental aspect of daily life, a better understanding may result in useful applications. That famous bitter medicine could be defused by addition of an anti-bitter agent (rather than trying to cover it up with a lot of sugar). Healthy food could be made to taste better than unhealthy stuff. If you don't like what you're eating, maybe you could retune the taste to something better. Most importantly, you could have some interesting dinner party conversation about how taste really works.

(2000)

Further Reading

J. Chandrashekar *et al.*, *Nature*, 2006, **444**, 288.

What Happened Next

In 2002, Charles Zuker's group reported a receptor protein for L-amino acids, i.e. the umami taste. It was found to consist of different protein molecules (T1R1 and T1R3), one of which it shares with the sweet receptor, T1R2:T1R3.

Further Reading

G. Nelson *et al.*, *Nature*, 2002, **416**, 199.
X. Li *et al.*, *Proc. Natl Acad. Sci. USA*, 2002, **99**, 4692.

Let Your Love Glow

I was intrigued by the finding that sexual arousal uses the same molecular messenger in the firefly (not exactly our nearest relative) and in the human male. I explored the possibilities in a humorous column in *Nachrichten*, but below is the version of the story where I tried hard to remain serious.

The love life of fireflies has been studied in great detail. Mainly for voyeuristic reasons, of course, as scientists enjoyed watching the male light up its lantern in a series of short flashes (a few hundred milliseconds) to attract the female, who then signals back to express her readiness. Starting with the classic works of William McElroy in the 1940s and 50s, biochemists have worked out the light reaction involving an enzyme (luciferase), its substrate (luciferin), and oxygen, and they identified a nerve signal that controls the flashes.

But, as the nerve endings are some 17 micrometers away from the cells which produce the light, it was unclear how the signal bridges this gap. Barry Trimmer and his coworkers at Tufts University (Medford, MA) have now found that for a fulfilling love life, the firefly depends on the same signaling molecule as the male of our own species: nitric oxide (NO).

Trimmer had previously studied the role of NO in caterpillars, but after hearing a talk on fireflies decided to check whether they use this molecule too. The fact that nitric oxide synthase is expressed in their lantern suggested they might. When his group locked up fireflies in an atmosphere containing some 70 ppm (parts per million) of this gas, they could observe the insects glowing permanently. In order to demonstrate that the NO is acting on the light organ rather than on the brain, they also prepared animals whose nerve connections to the

The Birds, the Bees and the Platypuses. Michael Gross
Copyright © 2008 WILEY-VCH Verlag GmbH & Co. KGaA, Weinheim
ISBN 978-3-527-32287-9

light organ were interrupted. While these were unable to flash spontaneously, the exposure to NO again resulted in rapid flashing or even permanent light. Conversely, when they stimulated the lantern by supplying the nerve signal, they could suppress the light emission by adding NO scavengers.

Thus there is clear evidence that NO is involved as a messenger in the final steps of the signaling. Quite how it fits into the chain of events remains to be clarified, but Trimmer speculates that it may inhibit the respiration process in the light-emitting cells, thus making available oxygen that would otherwise have been used up by the mitochondria. This is consistent with the observation that other inhibitors of respiration (such as cyanide) also induce the luminescence. It also suggests a mechanism for terminating the short flashes: As white light suppresses the effect of NO on respiration in mammals, the firefly's light might literally turn itself off.

Seeing that arousal in men is communicated by the very same NO (it activates the synthesis of cGMP, an effect which is enhanced by the drug Viagra which blocks the degradation of this compound), all we need to let our love glow is a lantern…

(2001)

Further Reading

Educational web page on the firefly project: http://ase.tufts.edu/biology/firefly/
B.A. Trimmer et al., Science, 2001, 292, 2486.

What Happened Next

NO has kept popping up in the most unlikely places. For example, reports published in October 2007 suggest that a main reason why generous blood transfusions can be dangerous to the patients concerned is the loss of nitric oxide from banked blood. Apparently, the oxygen carrier hemoglobin also carries nitric oxide, which serves as a signal for widening the blood vessels. Sound familiar?

Jacobson's Molecules

To continue on the theme of sexual signaling: In 2002, molecular biologists were homing in on the pheromone receptors responsible for the still largely mysterious function of Jacobson's organ in mice. Again, this is a supposedly serious version of a story which I've also done in a humorous format. I'm (seriously) looking forward to the day when human pheromones and their receptors will be brought to book.
PS: Watch out for the phrase I borrowed from Alanis Morissette's lyrics.

The idea that sub-conscious chemical sensing of pheromones could guide our sexual behavior is somewhat unsettling. So, at the end of the day, all the social games, intellectual intercourse, all-too-obvious or subtle signaling, clever scheming, and romantic thoughts we invest in the dating game might be overruled by a chemical signaling system that we share with mice, and even insects, and which is so primitive that its output never registers with the conscious brain? Outrageous!

Little surprise that it took a long time to make such ideas respectable. Even though the vomeronasal organ (VNO, also known as Jacobson's organ in honor of a researcher who published descriptions of animal VNOs in 1811) was discovered in humans nearly 300 years ago, its function in mammals and especially in humans has fuelled debates ever since. In the middle of the 20th century, anatomy textbooks claimed that the human VNO, if it could be found at all, was a non-functional vestige of evolution. Since the early 1990s, however, new evidence in favor of its function has accumulated.

The case for VNO function is more clearly established for reptiles and non-primate mammals, especially rodents. In the 1970s, it was

The Birds, the Bees and the Platypuses. Michael Gross
Copyright © 2008 WILEY-VCH Verlag GmbH & Co. KGaA, Weinheim
ISBN 978-3-527-32287-9

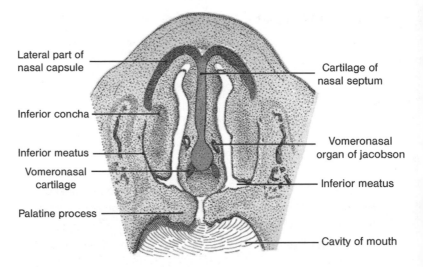

Figure 13 The vomeronasal organ of a human embryo.
From *Gray's Anatomy*, 1918.

shown that surgical removal of the organ in mice leads to severe impairment of sexual and social behavior without affecting the separate odor recognition behavior which is vital for recognition of food. Thus, the function of Jacobson's organ was finally established on a physiological level, but the molecules involved in its function remained elusive.

Pheromones act at extremely low concentrations and often in subtly balanced mixtures that are difficult to analyze. Add to this the fact that they don't talk to our conscious brain, and it becomes clear why we know so little about them. It was only after a systematic exploration of the mouse genome that molecular biologists could get a handle on some of the mammalian pheromone receptors, of which there are possibly more than one hundred.

In 2002, Richard Axel and his coworkers at Columbia University, New York, were the first to succeed in changing the sexual and social behavior of mice by removing the gene for a specific protein in the VNO. The protein they targeted was a cation channel – a protein located in the cell membrane that can allow positively charged ions to cross the membrane – known as trp2. The researchers created a strain of mutant mice that lack both copies of this gene (known as knockout mice), but appeared to be normal and healthy. Behavioral studies of the mutant mice revealed a number of unusual features. Male mutant mice were less aggressive towards intruders and a lot more likely to engage in homosexual activities. Female mutants were less aggressive in defending their offspring against intruders. The wide range of changes resulting from this single gene knockout suggests that this ion channel may have a general role in VNO signaling, rather than a specific role on just one pathway.

The group led by Peter Mombaerts at Rockefeller University, New York, reported a first step towards identification of the molecular receptors responsible for VNO function, along with the pheromones that might bind to them. Studies of the mouse genome have identified more than 100 candidate genes for VNO receptors, which are now collectively known as the V_1r superfamily. Mombaerts' group focused on a cluster of 16 of these found on chromosome 6, containing most members of two smaller groups, the subfamilies V_1ra and V_1rb. The researchers created mutant mice by deleting this entire gene cluster and studied their behavior.

As was to be expected, significant changes in sexual behavior were observed in this mutant, and they were characteristically different from those induced by both the trp2 knockout and by surgical removal of the VNO. One intriguing observation was that the receptor knockout mice (like the VNO-less ones) got increasingly bored with sex, while the wild-type and trp2 mutants became keener on it as their experience increased. This finding suggests that the VNO makes certain behaviors feel rewarding.

At this stage, the experiments mainly serve to establish that there is a clear link between genes expressed in the VNO and certain types of sexual and social behavior, at least in rodents. Many more studies addressing single genes, protein products, and pheromones, along with the interactions and signaling networks downstream of the pheromone receptors, will have to be carried out to clarify what exactly is going on in the "second nose."

And then there remains the important issue of whether all of this is true for humans too. Even though there are deodorants in the shops which allegedly address Jacobson's organ, it will be a long time before those inexplicable attractions will become explicable. But with the rodent receptors as a vital clue and the human genome data on file, it should be possible to track down the human VNO receptors and test those products on them. So if you're afraid of your animal instincts, you may be able to control them in the future. But life might become less interesting.

(2002)

Further Reading

L. Watson, *Jacobson's organ*, Allen Lane, 1999.
B.G. Leypold *et al.*, *Proc. Natl Acad. Sci. USA*, 2002, **99**, 6376.
K. Del Punta *et al.*, *Nature*, 2002, **419**, 70.

What Happened Next

Not much, I'm afraid. I'm still waiting for the human pheromone receptors to come out of their hiding places!

The Science of The Simpsons

Well, I don't think I'll have to explain why this one was fun to write. The idea occurred to me one morning while I was walking the Simpsons' fan in my family to school.

I never managed to see the point of TV series. Like church, they appeared to me as repetitive exercises for the faithful only, with relatively little enlightenment. But I had to revise my blanket condemnation of all things episodic when my youngest daughter got addicted to "The Simpsons" and I found myself increasingly drawn into the series as well. For her, the initial attraction was the yellowness of the characters – yellow being her favorite color. For me it must have been the humorous yet highly sophisticated angle the series offers on science.

It is abundantly clear that the people behind "The Simpsons" are proceeding with a scientific rationale. The core object of their investigation, the Simpson family, is a system trapped in a dynamic yet extremely stable equilibrium. In over 13 years and more than 300 episodes, virtually nothing has changed irreversibly. Even the children haven't grown older: Bart and Lisa are still in the same class at school, and Maggie is still sucking her dummy. While the non-ageing policy is widespread in cartoon world, only scientists would take as much care as the Simpsons team to ensure the constancy of the starting conditions in everything they do.

In every episode, the writers change just one parameter in order to probe the response of the equilibrium system. The change may temporarily affect many people in Springfield and turn their small world upside-down, but by the end of the episode, the system will have returned to the initial state. The unpredictably meandering path on

The Birds, the Bees and the Platypuses. Michael Gross
Copyright © 2008 WILEY-VCH Verlag GmbH & Co. KGaA, Weinheim
ISBN 978-3-527-32287-9

which it returns allows us to observe the mechanisms of reactions between the system's components. For example: Homer breaks his jaw and has to wear a brace that doesn't allow him to speak. To break out of the isolation, he encourages others to speak to him about their problems and he learns to listen. Thus he is suddenly seen as a thoughtful and understanding person by all around him. Other initial disturbances include Bart and Lisa being transferred to different classes or different schools, Marge rediscovering high-school admirers, and Grandpa Simpson falling in love. Each of these (and 300 other) experiments triggers major reactions, but you may rest assured that by the end of the episode, everything will have returned to the initial equilibrium state.

A notable, once-every-year exception to this rule are the Halloween episodes, forming a mini-series under the title "Little Treehouse of Horror." In total reversal of the general policy, these episodes feature "magical" and unrealistic events that snowball into ever bigger catastrophes, leading ever farther away from the normal state. While the normal episodes illustrate the "negative feedback" situation, where changes result in forces that lead back to the initial state, the Halloween episodes show positive feedback, where a small change can trigger a major catastrophe, and the planet is eventually taken over by dolphins, zombies, or aliens.

Circumstantial evidence for the scientific thinking behind the series is found in many science-based jokes, featuring the laws of thermodynamics, nuclear power, and evolution (often greatly egged on by radiation leaks from Homer's workplace). Modern technology is represented not only by the ever-present nuclear power station, but also in advances such as the notorious monorail. While there is no criticism of the technology as such, its failure is shown to result from the involvement of stupid operators (Homer S., responding to imminent meltdown by doing "eenie-meenie-minie-mo" to the control buttons of the power station), greedy proprietors (Mr. Burns) and gullible customers, like the people of Springfield who choose the pointless and flawed monorail project over real improvements of their infrastructure.

The lack of scientific knowledge in the general public is a recurring theme. Homer, of course, represents the absolute zero level of scientific literacy. His analphabetism in science matters is perhaps exposed most drastically when the family is playing scrabble and he moans:

"Nobody can make a word with these letters," but then we see the letters aligned in front of him: "O X I D I Z E." Even though his job at the power plant would in theory require some knowledge of physics, glimpses of understanding are extremely rare. Although Homer can be funny even when he defends scientific orthodoxy against Lisa, who is busy building a *perpetuum mobile* for a school project, shouting, "In this house we obey the laws of thermodynamics!" The acute dangers of ignorance are also personified in the rogue medic Dr Nick Riviera who doesn't know that things labeled "inflammable" can burn, and who operates on Homer's heart without having a clue as to which blood vessel connects to which chamber.

Lisa, in contrast, is the person to turn to for competent scientific answers on everything from astronomy to zoology. On some occasions, she even gets to practice real scientific research, for example when she isolates the pheromone that makes bullies attack nerds, and when she tests Bart's intelligence against that of a hamster. Her scientific prowess does her little good, however, as most of the other characters are too dumb to appreciate her knowledge. She is also lacking role models, as the only scientist to appear regularly is a "mad inventor" style nerd. Thus, while extending the scientific method of experimental analysis to the field of cartoon series, the Simpsons present some sobering lessons to real-life scientists.

But the good news is that this is a TV format where you can mention thermodynamics without scaring people away. As my kids and I are getting close to having seen all episodes, and there is the danger that "The Simpsons" may come to an end at some point, we desperately need more TV made by people who care about scientific understanding, not about blinding their viewers with techno-babble. In other words, give us less kryptonite and warp drive, and more power stations and three-eyed fish.

(2003)

Further Reading

P. Halpern, *What's Science ever done for us?* Wiley, 2007.

What Happened Next

On subsequent school walks, I pondered the option of expanding this topic into a book, following in the now well-established tradition of *The Physics of Star Trek* by Lawrence Krauss. I never got my act together to write this book, but Philadelphia-based physicist and science writer Paul Halpern did write such a book, which appeared in 2007 (on time for the Simpsons movie), see Further Reading, above.

Elusive Treasures

For a while, I thought about turning the following story into a book, reducing the all too colossal life and work of Fritz Haber to a manageable size by focusing on this rather eccentric and ultimately doomed sideline of his work. But then I shied away from the extra research involved in biography writing and limited myself to this brief sketch, which takes us back to the sexy topic of element 79, commonly known as gold.

If it comes to playing "six degrees of separation" with chemists, I happen to be reasonably well positioned. For example, the monumentally tragic figure of Nobel laureate Fritz Haber (1868–1934) is only three steps away, as my former PhD supervisor is a son of Haber's assistant and would-be biographer Johannes Jaenicke. I cherish this connection as it throws a somewhat unusual light on the great man.

The well-known parts of Haber's biography are all colossal in their impact on history and people's lives. Being a very patriotic German citizen during WW I, Haber helped to develop chemical weapons and oversaw their use at the front. Although aware of the horrendous suffering they caused, he reckoned they would bring a quick end to the war and thus reduce overall loss of lives. His young wife, Clara Immerwahr, one of the first women to gain a doctorate in chemistry, disagreed and committed suicide after failing to stop him. On the other side of the balance, one can estimate that roughly half the nitrogen feeding the world population today comes out of the Haber–Bosch process he invented. Bringing a complex biography to a tragic end, he had to flee Nazi Germany and died in Switzerland on the way to Rehovot, where Chaim Weizmann had offered him a position.

The Birds, the Bees and the Platypuses. Michael Gross
Copyright © 2008 WILEY-VCH Verlag GmbH & Co. KGaA, Weinheim
ISBN 978-3-527-32287-9

Haber's former assistant Johannes Jaenicke, together with his wife, spent decades collecting materials for a biography, which he never finished owing to the sheer size of the task and his failing eyesight in old age. (The existing biographies by Dietrich Stoltzenberg and Daniel Charles are largely based on Jaenicke's material.) One research project that Jaenicke was involved in himself took place between the World Wars, and offers a little light relief in the Faustian life of Haber. In the spring of 1920, Haber surprised Jaenicke and other coworkers with the announcement that he wanted to investigate the possibility of extracting the gold content of sea water, which was estimated to be 5–10 parts per billion (ppb). If that figure was correct, he reckoned, extraction of significant amounts of gold should be economically feasible and might help Germany pay back the crushing debts resulting from WW I and the Versailles treaty.

Haber set up a group led by Johannes Jaenicke to develop this project further. For five years, up to 12 research staff and PhD students worked in strict secrecy on the gold project. While industrial sponsors including Degussa were informed, the project was kept secret from the Allied authorities. Initially, the researchers worked on improving the analytical and separation techniques. While there was only a lim-

Figure 14 The team that tried to extract gold from sea water. Fritz Haber is standing in the middle, Johannes Jaenicke is standing on the right.

(From: Stoltzenberg, Dietrich: Fritz Haber. Chemiker, Nobelpreisträger, Deutscher Jude. Wiley-VCH Verlag GmbH, Weinheim 1998, S. 481)

ited number of sea water samples available, their analyses seemed to confirm that the gold content was in the parts-per-billion range. For the separation, they tested different approaches and eventually settled for binding the gold to colloidal sulfur and filtering with sand also charged with sulfur.

By the summer of 1923, the experiments were ready to be transferred onto an actual ocean-going ship. Still in strict secrecy, a group of researchers, including Haber himself, boarded the passenger vessel *Hansa* bound for New York, officially registered as crew members. Legend has it that Haber was greatly amused to be called up as a "supernumerary accountant." While the chemists worked behind closed doors, rumors among the passengers ran wild, and upon arrival in New York, a newspaper reported: "German scientists see way to drive ships by using mysterious force." A second expedition in the autumn of the same year took Haber's team to Argentina.

The samples analyzed during these two trips gave very different results, such that the researchers had to go back to the laboratory and improve their analytical methods even further, excluding both gold gains (from common chemicals, dust, or jewelry) and losses (e.g. from adsorption to equipment). Thousands of bottles were filled with sea water samples from all corners of the globe, stacked in purpose-built wooden chests, and shipped to Berlin for analysis. However, the disappointing result was that the actual gold content in most samples was two orders of magnitude lower than the early analyses had suggested, which ruled out any hope of economic retrieval. In 1927, Haber published the results and buried the project.

While it failed to reach the goal, however, the project did improve analytical methods, and also helped to secure the funding of Haber's research in difficult times. Even more ironically, if the project had delivered large quantities of gold, the devaluation of the precious metal might have backfired. As the Spanish found out after plundering the gold of Latin America's native population, "finding" tons of gold doesn't automatically make you rich in the long term. Nitrogen from the air was less glamorous, but ultimately more valuable.

(2003)

Further Reading

D. Stoltzenberg, *Fritz Haber: Chemist, Nobel Laureate, German, Jew*, Chemical Heritage Foundation, 2005.

D. Charles, *Master Mind: The Rise and fall of Fritz Haber, the Nobel Laureate Who Launched the Age of Chemical Warfare*, Ecco, 2005.

R. Hahn, *Fritz Habers (1868–1934) Forschungen zur Gewinnung von Gold aus Meerwasser*, Master's thesis, Technical University of Berlin, 1995.

What Happened Next

Mining the oceans is a topic that recurs at regular intervals in new and old guises. These days, people are mostly after elements such as manganese, but as the laboratory methods improve over time, I wouldn't be all that surprised if somebody revisited the old quest for ocean gold.

Eggs and Sperms and Rock'n'Roll

To this day, Britain seems to have a number of hang-ups around the issues of sex and drugs, which lead to measurable problems such as the record-breaking rate of teenage pregnancies, and an epidemic of binge drinking. All the more surprising, then, that the same country has played such a pioneering role in *in-vitro* fertilization (IVF). Hang on, maybe there is method in this madness. No sex please, we'd rather have IVF instead? In any case, IVF also gave Britain a head start when the whole cloning and stem cells debates came rolling in.

Bioethical questions surrounding human reproduction, embryos, and stem cells make front page news nearly every week. In quick succession, you might see news concerning the cloning of human embryos for the purpose of producing stem cells (therapeutic cloning), discussions about anonymity of sperm donors, sex selection, cloning claims from maverick scientists, IVF mix-ups, "designer" babies, etc.

These and similar issues affect most European countries in some way, although each has its own potentially explosive mixture of science, religion, and politics. In Germany, for example, memories of Nazi eugenics have led to highly restrictive legislation, and vivid public debates.

The UK is relatively lucky in that an authority to deal with these issues was already in place before they started to boil over. The Human Fertilisation and Embryology Authority (HFEA, www.hfea.gov.uk) was set up in August 1991, as required by a law passed the previous year. Its main tasks were to

The Birds, the Bees and the Platypuses. Michael Gross
Copyright © 2008 WILEY-VCH Verlag GmbH & Co. KGaA, Weinheim
ISBN 978-3-527-32287-9

- license and monitor IVF clinics
- license and monitor research on human embryos
- regulate the storage of gametes and embryos

In addition, it also has advisory roles to the government and to patients. While this brief must have looked innocent enough in 1991, it has since then grown to include a huge range of extremely sensitive bioethical issues.

The HFEA, based in East London near Liverpool Street Station, is both an authority consisting of members who meet once a month, and a government department staffed with civil servants and equipped with a budget of around two million pounds. The authority's board acts as a supervisory board to the department structure. Ruth Deech, a law professor and warden of St. Anne's College at Oxford, chaired the authority from its foundation through to the spring of 2002. Her successor is Suzi Leather who has a background in political science and consumer issues. She was the deputy chair of the Food Standards Agency from 2000 to 2002.

One of the biggest moral dilemmas facing the authority over the last few years was the saga of the permission to select embryos for implantation that might become life savers for their siblings. On a background of misrepresentation by the media, who coined the word "designer baby" even though there never was any creative design process involved, the authority ruled that the selection of an embryo based on pre-implantation genetic diagnosis (PGD) was allowed in the case where it helps to ensure the child resulting from the treatment would be free of inherited disease. If, in addition to being healthy, it would also be able to provide umbilical cord stem cells that could help to cure a sibling, even better. In contrast, the authority denied parents permission to use PGD solely for the benefit of the suffering sibling in cases where the new baby would not be at risk of genetic disease.

On these grounds, the authority refused permission for the family of little Charlie Whittaker to produce a tissue-matched sibling by PGD (which did not stop them, however, from having the procedure carried out in the United States). On the other hand, the HFEA approved the application from the family of Zain Hashmi. As Zain's blood disorder, beta thalassaemia, is genetic, PGD would not only help to produce a tissue-matched sibling but also ensure that the new baby would be free of the disease.

Public trust in the Authority's authority was badly shaken in December 2002, when the High Court ruled that the HFEA had no right to grant the Hashmis this permission. In April 2003, however, the Court of Appeal toppled that decision and ensured the Hashmis would be free to proceed regardless of the outcome of any further legal action. Currently, all such cases have to be decided by an HFEA committee on an individual basis, following the stringent criteria mentioned above.

In its role of supervising embryo research, the HFEA is also in charge of deciding who is allowed to produce human embryonic stem cells in the UK. So far it has only granted three licenses to generate new stem cell lines: One went to King's College, London, where the group of Stephen Minger successfully produced the first human ES cell line in the UK. The most recent license was granted in June 2003 to the Roslin institute near Edinburgh, famous for the cloning of Dolly the sheep.

While the early arrival of the HFEA on the bioethics scene means that it already had an established structure and reputation by the time things got complicated, it also means that the 1990 parliament act on which it is based could not possibly foresee most of the problems the authority is now faced with. At the HFEA's annual conference in January 2004, chairperson Suzi Leather launched a review of the old law, saying it had become "anachronistic" in places. For instance, she addressed the law's requirement for the IVF doctors to attend to the "need of a child for a father" before proceeding. The authority's review of the law is due to be presented to the government by the end of this year, but it looks likely that consultations will be held in 2005, so it may be some time before a new law could be passed. Like Alice racing the Red Queen, legislators have to run faster and faster to keep up with changes in the field of fertility and embryo research. In the absence of any perfect solution to this fundamental problem of fast-moving technologies, the HFEA, with its reputation for balanced decisions, looks like a really useful thing to have.

(2004)

Further Reading

www.hfea.gov.uk

What Happened Next

Amazingly, the HFEA survived the Blair premiership without being privatized or reorganized beyond recognition. As of October 2007, the chairperson is Ms Shirley Harrison, who also chairs the Human Tissue Authority (HTA). There appear to be plans to merge the two institutions into a combined Regulatory Authority for Tissues and Embryos (RATE). In September 2007, the HFEA ruled that research with hybrid embryos merged from different species, including human ones, can be permitted on a case-by-case basis. In a statement released in support of this decision, the authority said that "there is no fundamental reason to prevent cytoplasmic hybrid research." However, this decision did not include a specific approval of the two pending applications to permit such research. The authority was due to rule on the specific cases in November 2007.

A Cuban Success Story

In April 2004 I visited Cuba for the first time, and my Cuban friend, Reynaldo Villalonga, arranged around a dozen meetings with scientists at the Universities of Matanzas and Havana. One of the researchers I talked to was Vicente Verez-Bencomo, who told me the following story, which was in press at the time and came out a month later in *Science* magazine.

Children in Britain are routinely immunized against *Haemophilus influenzae* B (HiB), which causes meningitis and pneumonia, but in the developing world there are literally billions of them whose families cannot afford the vaccine. As a result, more than half a million children die from HiB infection each year. In 2004, Cuban researchers presented a novel, fully synthetic vaccine that can be produced much more cheaply and thus will reach all parts of the world. Coincidentally, it is the first fully synthetic vaccine to succeed in all clinical trials.

Over the last two decades, Cuba has responded to the lack of affordable drugs by investing heavily in biotechnology. As part of this effort, the laboratory of Vicente Verez-Bencomo at the University of Havana set out to produce an alternative to the original HiB vaccine, which Verez describes as "only a solution for rich countries." The key challenges were to produce the characteristic oligosaccharides found on the cell surface of the pathogen, and to find a way of presenting them to the immune system that would secure a strong, specific, and long-lasting immunity, i.e. the presence of antibodies that would be able to raise a rapid defense effort in case the real pathogen showed up.

The Birds, the Bees and the Platypuses. Michael Gross
Copyright © 2008 WILEY-VCH Verlag GmbH & Co. KGaA, Weinheim
ISBN 978-3-527-32287-9

In collaboration with Canadian chemist René Roy and colleagues from several of the major Cuban biotech research centers located in the Western suburbs of Havana (including the Centre for Genetic Engineering and Biotechnology), Verez's group developed a novel synthesis route to the crucial oligosaccharides, suitable for large-scale production. By the end of the 1990s, they could produce the sugars, couple them to a human serum protein, and show that they triggered the desired immune response. However, in order to compete with the long-term effect of the existing vaccine, the researchers had to couple the sugars to a protein that stimulated the immune system more efficiently. Tetanus toxoid (TT) protein turned out to be the right molecule for the job.

The synthetic glycoconjugate went through all the required toxicity tests and clinical trials, performing at least as well as the commercial vaccine. "We have transferred production to the Cuban Biotech companies," says Verez. "They are going to produce one million doses by the end of the year."

(2004)

Further Reading

V. Verez-Bencomo et al. Science, 2004, 305, 522.
http://www.myhero.com/myhero/hero.asp?hero=Hib_Vaccine_Tech_2005

What Happened Next

In November 2006, the World Health Organization released a new position paper on the HiB vaccination, stating that: "In view of their demonstrated safety and efficacy, HiB conjugate vaccines should be included in all routine infant immunization programs. Lack of local surveillance data should not delay the introduction of the vaccine, especially in countries where regional evidence indicates a high disease burden."

In September 2007, India launched a similar vaccine developed by the local biotech company Bharat Biotech.

The Birds, the Bees, and the Platypuses

I suggest a title with every piece I write, but editors tend to have minds of their own, so less than half of my suggested titles survive in the published versions. This one got through, and it must be my favorite title ever. But the story is sexy, too.

What is the molecular difference that makes us male or female? At first glance it's simple: the male has a (fairly degenerate, see page 38) Y chromosome instead of the second X chromosome. More precisely, the presence of a single gene in the Y chromosome, known as SRY, makes you develop male characteristics. In its absence, the development reverts to the default option, which is female. Things get more complicated when biologists start talking about the birds and the bees. In birds, you know, it's the other way round: females have a pair of different sex chromosomes, while males have a matched pair – and don't even think about the bees. But the record holder for the most confusing sex determination system must be the duck-billed platypus. After decades of uncertainty, Australian researchers have established that this animal has no less than five pairs of sex chromosomes, including one that resembles ours, and one that is more reminiscent of birds.

The ever-popular platypus is one of only three surviving species from the deepest branch of mammalian evolution, the monotremes. Thus, its sex determination is of interest not just as a curiosity but also for any light it might throw on the early evolution of our mammalian ancestors. Using fluorescence *in situ* hybridization (FISH), Frank Grützner's group at the Australian National University in Canberra has sorted out the platypus's ten sex chromosomes, which have the confusing habit of merging into one large chain during cell divi-

The Birds, the Bees and the Platypuses. Michael Gross
Copyright © 2008 WILEY-VCH Verlag GmbH & Co. KGaA, Weinheim
ISBN 978-3-527-32287-9

Figure 15 The duck-billed platypus (*Ornithorhynchus anatinus*) lives in the eastern coastal regions of Australia. As its evolutionary ancestry separated from the main branch of mammalian evolution quite early, it can provide interesting insights into the evolution of certain mammalian traits, such as the sex determination system described here.
(From: http://en.wikipedia.org/wiki/ Image: Platypus.jpg. Photo by Stefan Kraft)

sion. They found that there are five male-specific (Y) chromosomes, which can pair up with five different X chromosomes. In the chain, they are always found in the same order. At one end of the chain there is a pair that resembles our own XY pair (although it confusingly lacks the SRY gene), but the pair at the other end shows some similarity with the ZW chromosomes of birds. The authors even suspect that the latter pair was the first to develop a sex-specific difference, while the others were recruited later, and the one that resembles ours came in last.

This surprisingly bird-like feature in the mammal that lays eggs and sports a duck-like bill might overthrow the old dogma that sex chromosomes evolved independently in birds and mammals. Maybe we originally inherited the system still found in birds and morphed it into ours. The platypus may have preserved the transition state of this important evolutionary change.

(2004)

Further Reading

F. Grützner *et al.*, *Nature*, 2004, **432**, 913.

What Happened Next

I'm ashamed to admit that I haven't followed platypus's sex life as closely as I should have, so I don't really know. I'll make a New Year's resolution to catch up with this.

Our Hairy Cousins

I wonder why the common chimp is always cited as our closest living relative. Strictly speaking, the sex-obsessed pigmy chimp, or bonobo, is just as close to us. But given the puritan attitudes prevailing in some parts of the world that I could think of, it is not all that surprising that the clean-living common chimp is often seen as the more presentable relation, and that it got its genome sequence sorted out first. Hence, the following story is not quite as sexy as it could have been, but still ...

The bishop of Oxford wanted to have the family relations clarified once and for all. During the famous meeting of the British Association for the Advancement of Science, held in the newly built University Museum at Oxford, Samuel Wilberforce asked Thomas Huxley, whether he was descended from the apes via his grandfather's or via his grandmother's side. Huxley declared he would rather have an ape as a grandfather than a human like Wilberforce. At least that is what the legend of this event tells us. Spoilsport historians tend to point out that all accounts of this argument are from Huxley's supporters, so the presumed "victory" of the nascent pro-evolution camp may have been enhanced by the reporters.

Almost a century and a half later, scientists can address such questions calmly and in detail using genome sequencing. Thus, we can state with some confidence that the last common ancestor linking Thomas Huxley to Clint, the first chimp with a full genome sequence, must have lived some five million years ago.

Family trees aside, the chimp genome is extremely useful mainly because it serves as a helpful comparison and reference point in stud-

The Birds, the Bees and the Platypuses. Michael Gross
Copyright © 2008 WILEY-VCH Verlag GmbH & Co. KGaA, Weinheim
ISBN 978-3-527-32287-9

ies of the species which Jared Diamond has called "the third chimpanzee," namely *Homo sapiens*.

For instance, the simian DNA offers insights into the evolution of our genome in the last five million years, and into the evolution of individual genes and the selective pressures that guided it. Arguably the most important role of the chimp genome will be that of an external reference in studies of human diversity, in particular with respect to the different responses to disease and drugs (pharmacogenetics).

Family Tree

Let's start with the family tree issues, though, to get these burning questions out of the system. In 2002, Morris Goodman and coworkers concluded on the basis of partial genome information from chimps and other primates that the genus *Homo* (setting us and the extinct hominids such as Neanderthal apart from chimps) has no justification in genetics. The traditional division dating from 1963, in which chimps are closer to gorillas than to humans, had been made obsolete by genome research. Thus, Goodman suggested opening up the genus *Homo* to include both the common chimp (*Pan troglodytes*) and the bonobo (*Pan paniscus*).

In 2005, the Chimpanzee Sequencing and Analysis Consortium published a 94% complete draft version of the chimp genome and a first preliminary comparison with the human one, adding further evidence in support of a very close family relation. Looking at differences between specific DNA bases ("letters"), the researchers found that 1.23% of them differ between humans and chimps. However, around 1/7 of these are probably due to natural diversity within each species, which leaves only an estimated 1.06% of real differences.

This figure refers to those 2.4 billion DNA bases of the chimp which researchers have been able to align (place side by side for comparison) with the corresponding DNA from the human genome. Thus, this material contains some 25.4 million small differences between us and our cousins. Within our genome, there are around 4 million bases that differ between individuals (referred to as SNPs, or single nucleotide polymorphisms). In the initial analysis, the researchers have found 1.66 million SNPs in the chimp genome.

Replacements of single letters aren't the only genetic differences that have driven us apart from the chimps over millions of years. On top of that, there is also a small number of insertions and deletions of longer stretches of DNA, collectively known as "indels," as well as rearrangements. The genome researchers have estimated that five million indels have provided each of the two species with some 40 to 45 million bases that aren't found in the other. Altogether, the different kinds of differences add up to some 3% of genetic deviation between chimps and humans. Alternatively, if you come from the "glass is half-full" school of thinking, you could say that genetically we're 97% identical with our hairy cousins.

But should this high degree of similarity be an excuse for us to go bananas or behave like animals? Comparative genomics in other parts of the animal kingdom have shown that percentage identity is a useful indicator of evolutionary history, but that it isn't necessarily a measure of similarity between species.

The chimp researchers use the example of two species of mice that are genetically just as different as humans and chimps, but which look pretty much the same from the outside. Dog breeds represent the opposite extreme. Using just 0.15% of genetic variability, breeders can produce spectacularly different looks.

Figure 16 The common chimp (*Pan troglodytes*) and the bonobo or pigmy chimp (*Pan paniscus*) are our closest living relatives. The investigation of the chimp genome has provided important insights into the evolutionary history and current genetic diversity of our own species.
(© Digital Vision)

The cautionary take-home lesson is that appearances depend on a small number of genes. This is also the reason why the concept of human "races" is practically meaningless in genetics. There is much more genetic variability to be found between Africans (all supposedly belonging to the black race) than in all other human races taken together.

And even in the more hidden qualities that we human beings may take pride in, a few genes can have disproportionate influence. One notorious Achilles heel is a group of regulatory proteins known as transcription factors – they control the transcription of DNA into messenger RNA and thus the activity of many other genes. A single mutation of such a factor can easily derail metabolism, embryonal development, or the cell division cycle. For instance, mutations in the transcription factor p53 are associated with a large fraction of human cancers.

Thus it would be naive to assign the obvious and not-so-obvious differences between tree-dwelling apes and those that drive cars or write popular science books to this 3% difference. Only a much more detailed understanding of the function of the entire genome in embryonal development will enable us to understand the small difference between our closely related species.

Comparing the Genes

Let us now zoom into the genomes to find out what the chimp and other animals can teach us about our genes. The consortium that analyzed the chimp genome selected 13,454 genes that can be directly compared between humans and chimps. The researchers also identified a smaller set of 7043 genes that can be compared between humans, chimps, rats, and mice.

In order to judge the speed with which beneficial mutations are selected and deleterious ones blocked out, researchers typically compare the rate of base exchanges that lead to a difference in the amino acid coded for with those that don't (as is often the case for the third base in a three-letter codon). The idea is that the "silent" or synonymous base changes are supposed to indicate the natural mutation rate, while the sense-changing mutations are subject to natural selection.

Researchers compare the relative frequency of both types of mutations. If in a given gene or region "meaningful" mutations are more frequent than silent ones, this suggests that natural selection has favored certain traits. In the opposite case, selection has been busy keeping the gene intact and protecting it from damaging mutations.

Applying this type of analysis to the whole human genome in comparison to the common ancestor shared by humans and chimps, the researchers found that the non-silent mutations are reduced by 78% with respect to the silent ones. Thus, roughly three quarters of non-silent mutations were damaging enough to be suppressed by natural selection.

Still, some damaging mutations managed to establish themselves and are now known as genetic disorders. Based on the genome comparisons, the researchers estimated that a quarter of the surviving non-silent mutations have some negative effects. (Their survival may be due to the effect setting in late in life, or, as in the classic textbook example of sickle cell anemia, due to positive side effects that benefit the mutation carriers.)

The four-way genome comparison that also included rat and mouse showed that the rodents have experienced much more selective pressure than the primates.

Vox Populi

Insights from the chimp genome also benefit an area of research that has absolutely nothing to do with apes or any other animal species: human population genetics. After decoding the human genome, researchers have started to address the differences between individuals and between ethnic groups. The main motivation lies in the fact that the effectiveness and tolerance of medicines often depends on the genetic make-up of the individual patient. Pharmacogenetics is a scientific discipline that aims at predicting which patient will benefit from which drug.

These efforts have already resulted in the identification of more than seven million human SNPs (single letter variabilities). However, the finding that there are differences between individuals doesn't tell us the story of how they came about. Which is the "original" version and which is the mutation? And how did the mutation spread? Such questions cannot be answered with the human genome alone, but with the chimp as a reference point, they can.

With the help of Clint's genome, the researchers have already been able to clarify the origins of 80% of the known human SNPs. In the cases that remain ambiguous (as the chimp base may be variable, or different from the human variants), additional primate genomes will hopefully clear up the matter.

(2005)

Further Reading

The Chimpanzee Sequencing and Analysis Consortium, *Nature* 2005, **437**, 69–87. (free access via *Nature*'s web focus:
   http://www.nature.com/nature/focus/chimpgenome/index.html)

What Happened Next

As of 2007, researchers are still busy filling the last gaps in the chimp genome, identifying the genes, and working out what they are for. You can watch the work in progress here: http://www.ncbi.nlm. nih.gov/genome/guide/chimp/ In the efforts to improve our understanding of our own genome with the help of related species, the ongoing sequencing of the Neanderthal (see page 70) is of similar importance to the chimp project.

Cupid's Chemistry

In February 2006, I was famous for a day, making media appearances around the globe. This was all due to a press release
which the Royal Society of Chemistry had issued ahead of Valentine's Day, based on the following story. I have to say, however,
that I was quite glad when Valentine's Day was over and the ra-
dio stations stopped calling.

Cupid, the mischievous little archer, may be all around us at this time
of the year, but there is little scientific evidence to support the age-old
claim that his arrows make people fall in love. Plato's beautiful explanation involving the loss of an "other half" wouldn't withstand today's
peer review either. And if anybody tries to sell you a love potion à la
Tristan and Isolde, you should not expect any miracles from it.

Despite the failure of the romantic explanations, the romantic phenomenon persists, and according to love researcher Helen Fisher it is
"a universal or near-universal cultural constant." There is no human
culture on Earth, she claims, that has been proven not to know the
phenomenon of romantic love.

If it is universal, scientists argue, there must be biological basis for
it. In other words, it cannot be simply a cultural tradition like cricket
or opera. In recent years, some researchers have boldly forsaken their
natural fear of the irrational side of the human being and set out to investigate the biological and chemical processes underlying romantic
love. In particular, they have studied the action of genes, neurons, and
chemical messengers such as hormones and pheromones.

The Birds, the Bees and the Platypuses. Michael Gross
Copyright © 2008 WILEY-VCH Verlag GmbH & Co. KGaA, Weinheim
ISBN 978-3-527-32287-9

Figure 17 Cupidon. Painting by William Adolphe Bouguereau (1825–1905)

Vole Story

Naturally, if some kind of biological phenomenon is universally found throughout a species, one would suspect it to be engrained in the genes in some form or shape. The trouble with love is that it is a complex phenomenon, presumably controlled by complex interactions between many different gene products. Thus, it would be difficult to study for the same reasons that apply to multifactorial diseases such as heart disease. Furthermore, the trouble with human subjects is that ethical concerns rule out manipulations of their genes, which would be required to deconvolute the interactions of many genes.

Therefore, genetic studies of mating and courtship have so far remained limited to animals and to relatively simple questions. The most spectacular and widely reported study of this type was conducted with two species of North American voles, namely the monogamous prairie voles (*Microtus ochrogaster*), and the genetically related montane voles (*Microtus montanus*), which do not form any bonds but copulate at random. Thomas Insel and Larry Young at Emory University in Atlanta, Georgia, discovered an insertion in a gene of the monogamous prairie vole which is suspiciously absent in the polygamous montane vole.

To test whether this insert is linked to the difference in sexual behavior, the researchers incorporated the gene with the insert into the genomes of male montane voles. Indeed, they succeeded in "curing" these rodents from their promiscuity with this simple genetic manipulation.

More recently, another "sex gene" has been tracked down in the fruit fly *Drosophila*. Ken-Ichi Kimura and coworkers at Hokkaido University demonstrated that the protein encoded by the *fruitless* gene of *Drosophila* controls the construction of a male-specific neural circuit which is thought to play a key role in male courtship behavior. Which neatly shifts our attention from genes to neurons and the brain.

Truly Madly Deeply

Modern brain imaging techniques such as functional magnetic resonance imaging (fMRI) or magneto-encephalographic (MEG) scanning are far from being just another tool in the box. They have opened up a whole new world of possibilities, because they enable researchers to observe the working brain without harming the patient.

Helen Fisher, an anthropologist at Rutgers University, joined forces with the New York researchers Arthur Aron and Lucy Brown to investigate the manifestations of early-stage romantic love in the brain. Essentially, they set out to establish whether love works like a fundamental emotion (e.g. fear) or whether it is produced by the feedback loops of the brain's reward system (like cocaine addiction).

The researchers recruited ten women and seven men who said to have been intensely in love between one and 17 months, and assessed them by interviews before and after the fMRI study. During the imaging experiment, each participant was shown a photo of their romantic partner and asked to recall any cherished memories linked to that person. As negative controls, they were also shown photos of other friends and family members and asked the same question. To purge any romantic feelings between the photos, the subjects were made to perform mental arithmetic, counting backwards from a randomly selected 4-digit number in steps of 7. (Try this procedure if you ever need to clear your brain of romantic overload – it seems to have been efficient within less than a minute!)

Comparing the brain scans of their subjects wallowing in romantic memories to those linked to neutral photos and those collected during the mental arithmetic exercise, the researchers were able to pin down several key regions of the brain that appear to be involved in intense romantic feelings but not, for example, in face recognition. Specifically, they recorded activation in the right ventral midbrain, around the so-called ventral tegmental area (VTA) and the dorsal caudate body and caudate tail. All these regions are unrelated to primeval instincts and emotions such as fear, but they are linked to the reward system that can get us addicted to drugs.

Reviewing their work in comparison with related papers, Fisher, Aron, and Brown conclude that "romantic love is primarily a reward system, which leads to various emotions, rather than a specific emotion." Characteristically, there is no facial expression that can be unequivocally linked to being in love. They also observe that early stage, intense romantic love is different from both the sex drive and the development of attachment in the later phases of a relationship, which activate different areas of the brain.

In a follow-up study, Fisher and her colleagues have started to look at what happens when love goes wrong. "We all get 'dumped' at one point or another," Fisher says. "So I wanted to see what happens in the brain when you are rejected in love." Accordingly, she and her colleagues applied the brain imaging technology to a group of 15 volunteers who had recently been dumped. From the preliminary results, Fisher concludes that "a lot happens in the brain when you look at a photo of someone who has just abandoned you, including activity in brain regions associated with physical pain, obsessive/compulsive behaviors, controlling anger, and regions that we use when we are trying to speculate on what someone else is thinking." Far from switching off the brain activities involved in the previous romantic bliss, Fisher finds that "it also appears that when you get dumped you start to love your rejecting partner even harder."

A key feature of the brain areas that the US researchers have connected to romantic love is that they are involved in signaling pathways using the hormone dopamine. But which other hormones can be blamed for the emotional rollercoasters of romantic love?

Molecules in Love

Donatella Marazziti, a psychiatrist at the University of Pisa, started out investigating the hormonal changes connected to obsessive/compulsive disorder, and then moved on to those that occur when people fall in love. Initially, she and her coworkers found a decrease of the functionality of serotonin transporters in the blood of enamored volunteers, who had been selected and rated on the "Passionate Love Scale" (PLS) much like those in the US studies above. Like obsessive/compulsive patients, the love-struck volunteers showed a reduced concentration of serotonin in the blood, which might explain why early phase romantic love can turn into obsession.

In her most recent study, Marazziti, together with Domenico Canale, cast the net wider to check for changes in the concentration of a number of hormones, including estradiol, progesterone, DHEAS (dehydroepiandrosterone), and androstenedione, which were found to be unaffected by any romantic feelings. In contrast, they observed changes for cortisol, FSH (follicle stimulating hormone), and testosterone. Some effects were gender-specific. For example, testosterone was found to be increased in women but reduced in men when they are in love.

If lovers swear their feelings to be everlasting, the hormones clearly tell a different story. Re-testing the same subjects 12–24 months later, Marazziti and Canale found that the hormonal differences had disappeared entirely, even if the relationships remained intact.

Using the same method for volunteer selection, Enzo Emanuele and his coworkers at the University of Pavia investigated whether a different class of chemical messengers, the neurotrophins, is involved in the romantic experience. They reported at the end of 2005 that the concentration of nerve growth factor (NGF) in the blood exceeds normal levels in enamored volunteers, and that it increases with the intensity of romantic feelings as measured by the PLS. Whether more NGF is needed in early stage romance because of all the new experiences that are engraved into the brain, or whether it has a second, as yet unknown, function in the chemistry of love remains to be explored.

Emanuele and coworkers, too, found that after 12–24 months all the love molecules had gone, even if the relationship survived. Neither the initial intensity on the PLS nor the concentration of NGF appeared to be a suitable predictor for the fate of the relationship after that period.

Another molecular messenger of love is phenylethylamine, a neurotransmitter that is structurally related to amphetamines. "Phenylethylamine is responsible for 'love at first sight,'" says Gabi Fröböse, who co-authored the book *Lust and Love – is it more than chemistry?* with her husband Rolf Fröböse. "But after the initial euphoria which may last two to three years, its effect fades."

But, if all the chemical messengers of intensive romantic feelings disappear within two years, what is the chemical glue that keeps (at least some) couples together?

A key molecule for the attachment phase is the hormone oxytocin, a nonapeptide that was first described as the chemical principle that induces labor and lactation, but later found a second job as the human "cuddle hormone." It is related to the hormone vasopressin, which controls kidney function and is also involved in the attachment of the above-mentioned prairie voles.

Experiments have shown that – depending on the species – either or both of these hormones can make animals snuggle up. In humans, it has been shown that oxytocin production is high during female orgasm, accounting for her desire for cuddles after the event. Apart from that, and its role in childbirth, very little was known about oxytocin's role in human physiology and psychology until very recently.

In 2005, several groups reported progress in the investigation of the role of oxytocin in humans, linking the hormone to early socialization, social cognition, and trust. Michael Kosfeld and his coworkers at the University of Zurich, in particular, showed that application of oxytocin via a nasal spray made participants in a "trust game" they devised more trusting towards other human participants, but not towards a computer. This finding fits in with the expectations of the Italian researchers. "I am not surprised by the results of Kosfeld's paper," says Donatella Marazziti, who has just completed a study of oxytocin in romantic love but keeps the details under wraps.

Cupid's Arrows

Finally, another family of chemical messengers associated with love, the pheromones, is equally poorly understood in humans, as most of our knowledge derives from animals. By definition, pheromones are chemicals intended for communication between individuals of the same species. Their use in insects is well understood to the extent that "pheromone traps" are commercially available for crop protection.

Our knowledge is much more incomplete for mammals, let alone humans. Most people's educated guess is that pheromones secreted from some glands, e.g. with the sweat, are recognized by receptors presumably located in that very small part of our nose known as Jacobson's organ or vomeronasal organ (VMO, see page 119). However, it was only in 2002 that researchers could pin down some putative mammalian pheromone receptors in mice. In October 2005, the group led by Hiroko Kimoto at the University of Tokyo added a surprising piece to the jigsaw. The Japanese researchers showed that a non-volatile mouse pheromone, which they called ESP-1 (exocrine-gland secreting peptide), is released from the tear glands of the male mouse and – after face-to-face contact – activates receptors in the VMO of the female.

Again, it remains unclear whether the tears of the human male have a similar effect. Indeed, there is a long-running controversy as to whether the human VMO is in fact a working part of our physiology or whether it's an inactive relic of mammalian evolution. It now appears that the evidence is slowly giving the pro-VMO party the upper hand. To any romantically inclined chemist, it should be deeply satisfying to be able to prove that chemical messengers communicate romantic feeling between humans. After all, this is the only thing that science can offer as a real-world analogy to Cupid's arrows.

(2006)

Further Reading

G. Fröböse and R. Fröböse, *Lust and Love – is it more than chemistry?* Royal Society of Chemistry, 2006.

H. Fisher, *Why we love – the nature and chemistry of romantic love*, Henry Holt, 2004.

D. Marazziti and D. Canale, *Psychoneuroendocrinology*, 2004, **29**, 931.

K.-I. Kimura *et al.*, *Nature*, 2005, **438**, 229.

A. Aron *et al.*, *J. Neurophysiol.*, 2005, **94**, 327.

H. Fisher *et al.*, *J. Comp. Neurol.*, 2005, **493**, 58.

E. Emanuele *et al.*, *Psychoneuroendocrinology*, 2005, **30**, 1017.

A. B. Wismer Fries *et al.*, *Proc. Natl. Acad. Sci USA*, 2005, **102**, 17237.

P. Kirsch *et al.*, *J. Neurosci.*, 2005, **25**, 11489.

M. Kosfeld *et al.*, *Nature* 2005, **435**, 673.
H. Kimoto *et al.*, *Nature* 2005, **437**, 898.

What Happened Next

No major advance has come to my attention, but if anything does turn up, I will of course have to wait for Valentine's Day to come round, because "seasonal" stories will get the biggest coverage in the media.

Colombia after Columbus

The sexiest people on this planet come, of course, from Latin America. What is less clear is in which ways exactly the genetic melting pot following the *conquista* brewed up this characteristic mix.

The population of Latin America is a rich and varied mixture of Indigenous, European, and African genetic heritage. The economic development in the New World during colonial rule followed essentially two different patterns. In the resource-rich tropical and subtropical areas, European slave drivers exploited cheap labor recruited by force either locally or in Africa. In the moderate climates of North and South America, however, European settler families drove the Indigenous population off the land.

These different approaches had drastically different effects on the genetic fate of the populations. In the tropical model, immigrants and natives eventually mixed and mingled in spite of racial prejudice, while the European settler families in the moderate climates kept their gene pool to themselves.

Thus, it is reasonably clear why most Canadians look different from most Mexicans, but until recently, there was very little scientific evidence to clarify how the broad and smooth spectrum of mixed heritage in the tropical and subtropical parts of the New World came about. First attempts at using genetic methods to solve this problem delivered contradictory results.

The group led by Andrés Ruiz-Linares at University College London, for instance, used gene sequencing to study the origins of the current population of the Antioquia region in the western part of Colombia (around the city of Medellín). First studies suggested that

The Birds, the Bees and the Platypuses. Michael Gross
Copyright © 2008 WILEY-VCH Verlag GmbH & Co. KGaA, Weinheim
ISBN 978-3-527-32287-9

the Colombians in that area, who see themselves as people of European origin, are in fact 94% European according to their Y chromosomes, which are only passed on in the paternal line. However, the mitochondrial genes, which are only passed on in the maternal line, told an entirely different story. In the female line of descent, today's Antioquians are 90% Indigenous, 8% African, and only 2% European.

Intrigued by this apparent paradox, Andrés Ruiz-Linares and his coworkers carried out further, more detailed, studies in the same region. In their improved analysis, they also included the variability of the X chromosome, and the abundance of common family names, analyzed in correlation with characteristic traits of the Y chromosome.

This detailed analysis provided much deeper insights into the history of the population. For instance, the researchers could show that five of the most common family names in the region can each be traced back to a single Spanish immigrant who arrived around the middle of the seventeenth century. Among the family names that were abundant from the very beginning, we also find Aristizabal, which lovers of Latin pop/rock music will recognize as the last name of Juan Esteban Aristizabal, or Juanes for short, who is in fact a native of Medellín.

After the arrival of the Spanish founding fathers, the genetic drama appears to have unfolded as follows. European adventurers who typically arrived unaccompanied, got involved with Indigenous women (of course the genetic research doesn't tell us whether the women consented or not!), creating the first generation of mestizos.

In the following generation, however, with mestiza women on the dating market, arriving men seem to have favored the half-bloods over the native girls, as the genetic contribution of the Indigenous population rapidly decreases to near zero. The researchers speculate that both a drastic decline in the native population and the (somewhat racist) preference of the European men may have played a role in this pattern.

The pattern continued over centuries in relative isolation, with males from European (paternal) lineage sitting at the top of the power hierarchy, and choosing (in the absence of European women) the whitest mestizas they could find. This behavior neatly explains why the Y chromosomes of the current population (along with the names) are mainly European, while the mitochondrial genes are mainly Indigenous.

Conversely, European women did not have much input, as the male invaders typically traveled unaccompanied, and Indigenous dads were losing out because their daughters were attractive only to the first generation of male immigrants. Just how this fractionated distillation of genes ended up producing the music of Juanes still remains to be investigated.

(2006)

Further Reading

L. G. Carvajal-Carmona *et al.*, *Hum. Genet.*, 2003, 112, 534.
S. Miller and J. Diamond, *Nature* 2006, 441, 411.
G. Bedoya *et al.*, *Proc. Natl Acad. Sci. USA*, 2006, 103, 7234.

Cheers to the Wine Genome

In the sexy part of this book, I have given you the bling, the lights, the sweet talking, and sexy people, so all we need now is something nice to drink. How about a glass of Pinot Noir, complete with insights into plant evolution and health benefits?

A consortium of researchers in France and Italy has completed a draft sequence of the genome of the grapevine *Vitis vinifera*, which yields valuable insights into the evolution of plants and into the specific abilities of the grapevine to produce a variety of aromas and a number of compounds associated with health benefits.

The researchers selected a grapevine variant derived from Pinot Noir cultivars, which were bred for high homozygosity, meaning that for most genes, both copies of the relevant chromosome carry the same version of the gene. Most grapevine variants have remarkably high number (up to 13%) of differences between the two copies, which would make the established method of shotgun sequencing extremely difficult.

With the selected variant, the consortium applied standard genome sequencing approaches and obtained more than eight-fold coverage of the complete genome. As this is only the fourth genome sequence of a flowering plant (after *Arabidopsis*, rice, and poplar), the investigators have been able to analyze the evolution of plant genomes at a greater depth than was previously possible. In particular, they were able to pin down evolutionary events in which whole plant genomes were duplicated or triplicated and locate them within an elementary "family tree" of these four species.

Moreover, the first analyses of the genome also reveal some of the traits that make wine special. For instance, the stilbene synthases

The Birds, the Bees and the Platypuses. Michael Gross
Copyright © 2008 WILEY-VCH Verlag GmbH & Co. KGaA, Weinheim
ISBN 978-3-527-32287-9

(STSs) that are responsible for the synthesis of resveratrol, a phenol derivative that has been linked to the health benefits of moderate wine consumption, are represented by a hugely inflated number of genes. The researchers found 43 STS genes, of which at least 20 are known to be active.

Anne-Françoise Adam-Blondon, who worked on the project at the Plant Genomics Research Unit at Evry, just south of Paris, thinks that applications will soon materialize. "It is now possible, for instance by a whole genome scan for linkage disequilibrium to relate specific haplotypes to wine characteristics, as well as to important agronomic traits such as pathogen resistance," she commented. Application research could be aimed "at the development of high quality grapevine cultivars resistant to diseases, and therefore will contribute to the much needed reduction of fungicide and pesticide treatments and to the development of sustainable viticulture conditions. Another track for application is also the development of cultivars producing [a] high level of health beneficial compounds for fruit juice production."

And what about the taste? The enzymes associated with aromas show an expanded presence in the genome compared with other plants. Terpene synthases (TPSs) are involved in a plant's interaction with its environment. But in the grapevine, they are also responsible for aromas. The other three plants that have been sequenced have between 30 and 40 such genes; the grapevine boasts 89 functional TPS genes and 27 pseudogenes. Thus the rich variety of flavors found in wines appears to be deeply rooted in the genomics of the grapevine.

(2007)

Further Reading

The French–Italian Public Consortium for Grapevine Genome Characterization, *Nature*, 2007, **449**, 463.
On resveratrol and lifespan (of mice):
J.A. Bauer *et al.*, *Nature*, 2006, **444**, 337.

3
Cool Technology

"Technology is just our word for stuff that doesn't work yet."

Douglas Adams

For parents of teenagers, "cool" is the most impenetrable adjective of them all, as the range of things that are or aren't cool seems to be changing randomly every day. However, one important quality of cool people is that they don't care what other people think of their taste. So here are a few things that I found really cool, but if the cool kids at your school find them nerdy, I'm totally cool with that.

Just to be on the safe side, some of the work included below deals with the indisputable meaning of "cool," namely low temperatures. Also, while much of my writing is in fundamental research, some of the cool stuff below is on the applied side of things, hopefully contributing to the cool technologies of the future.

Life on the Rocks

Let us, once more, be literally minded and start this section with something cool that allows cool creatures to survive at low temperatures.

Liquid water is one of the most important requirements for life. Deep-sea microbes can only thrive at 110 °C because the pressure of the water column above them increases the boiling point of water far beyond this temperature. While organisms at the upper limits of the biological temperature scale receive a little help from physical conditions, many inhabitants of ice-cold areas can actively keep water in the liquid state essential for their survival.

From everyday experience, we know various methods to prevent damage by freezing. Motorists, for instance, add antifreeze agents (such as the alcohol ethylene glycol) to the cooling and windscreen-wiping water of their vehicles; icy roads can be defrosted by salt. Both substances are very easily soluble in water, but cannot be included in ice crystals. Therefore, they pull the equilibrium between the liquid and the solid state of water in favor of the liquid and thus lower the freezing point by a few degrees.

Fish living in arctic waters cannot afford the luxury of a constantly high body temperature that we warm-blooded animals enjoy, but the motorist's method of mixing high concentrations of small molecules into their body liquids would get them into trouble as well. The natural tendency to equal concentrations (osmosis) would draw water into their cells and generate an osmotic pressure that could damage them at least as much as freezing and thawing. Therefore, various species of fish have evolved anti-freeze proteins (AFP), which interact specifically with the very smallest solid associations of water mole-

The Birds, the Bees and the Platypuses. Michael Gross
Copyright © 2008 WILEY-VCH Verlag GmbH & Co. KGaA, Weinheim
ISBN 978-3-527-32287-9

cules that start the freezing process, known as crystallization nuclei. By binding these nuclei, AFPs stop them from growing into larger ice crystals, effectively lowering the freezing point in their immediate environment.

In contrast, certain species of frogs and turtles do exactly the opposite. They avoid damage by facilitating freezing of their body liquids using ice nucleation proteins, which promote the formation of crystallization nuclei. The rapid freezing simultaneously starting from many such nuclei keeps the ice crystals so small they cannot cause any mechanical damage. And even bacteria possess – in addition to the well-studied heat shock proteins – a set of cold shock proteins to help them buffer the consequences of a temperature drop.

So we are looking at specific interactions of biomolecules with ice nuclei that have only just started to form in order either to stop them from growing or to facilitate their formation and growth – which sounds like a tricky problem requiring elaborate molecular structures. However, when the first crystal structure of such a protein was solved, it revealed a stunningly simple design. The antifreeze protein of the winter flounder (*Pseudopleuronectes americanus*), consisting of only 37 amino acid residues, is completely wound up to form a single alpha helix with nine turns. Researchers were surprised to find that such a simple structure, lacking the water-excluding core that is thought to be important for protein structures, can be stable and functional. They attribute the unusual stability of this single helix to special, cap-like structures linked by hydrogen bonds at both ends of the screw.

It is equally amazing to note that the protein contains only nine of the 20 different sorts of amino acid building blocks available. When protein experts look at the sequence, it isn't very hard for them to guess which amino acid residues contribute to the ice-binding function, as alanines – unsuitable for this purpose – take more than half of the positions. Researchers believe that all 14 non-alanine residues have important roles either in helix stabilization or in the recognition and binding of ice crystals. The ice-binding motif was identified as a triad of the amino acids asparagine, threonine, and leucine, recurring in every third turn of the helix. The same regularity is also found in bigger helical AFPs, with up to five repetitions of this motif. These residues form a surprisingly flat surface, from which the side chains of asparagine and threonine protrude only slightly. These can form hydrogen bonds with periodically recurring structures on certain surfaces of ice crystals, thus keeping the crystals from growing.

In contrast to the group of helical antifreeze proteins (type I AFPs) with their surprising simplicity, the type III AFPs found in other ocean fish, such as haddock, are somewhat more complex in their structures. In spite of their relatively small size, they have a quite intricately folded structure with no obvious periodicity or any other hints of ice-binding motifs. After researchers had solved the structure of a type III AFP by nuclear magnetic resonance (NMR) spectroscopy, they still needed tedious investigations exchanging individual amino acids one by one, to find out which ones are essential for the interaction with ice crystals. Meanwhile, it appears that out of the eight sequence strands involved in formation of beta-pleated sheet structures, the one nearest to the end of the sequence (C-terminus) forms the binding site.

Researchers are still all at sea regarding the type II antifreeze proteins, which are found in herring, for instance. Concerning these proteins there are no real structural data yet, only hypothetical structural models based on a rather remote kinship with a group of plant proteins, the lectins.

Not very much more is known about the ice nucleation proteins whose interaction with ice crystallization nuclei has the opposite effect, namely to facilitate nucleation and rapid freezing. This effect enables some species of frogs and turtles to survive freezing unharmed, although up to 65 % of their body's water content may turn into ice. Proteins with similar functions are secreted by certain microorganisms, specifically from the genera *Pseudomonas*, *Xanthomonas*, and *Erwinia*, in order to control ice formation in their immediate environment. As these proteins contain frequent repeats of certain sequence motifs, scientists suspect that a regular structure, possibly a large beta-pleated sheet, matches the periodicity of ice crystals.

Of course, the frogs may still find it stressful to be frozen, and there is some evidence that they cope with this by synthesizing specific cold shock proteins, in analogy to the heat shock proteins produced by most organisms upon exposure to higher temperatures.

Researchers hope to obtain more insight into the function of cold shock proteins by studying bacteria, which tend to be simpler in such fundamental things. Thus, the major cold shock protein of *E. coli*, CspA, was only identified and characterized in 1990, but its crystal structure was solved by 1994. It is virtually identical with the structure of CspB from *Bacillus subtilis* determined one year earlier, indicating that this function is evolutionarily quite old and predates the split of the bacterial kingdom into gram-positive and gram-negative bacteria. Both contain structures typical for proteins binding to nucleic acids (there is a homology with the ribosomal protein S1, for instance). Although their function has not yet been elucidated, they are most probably not directly concerned with the physical effects of the cold (as the AFPs are), but rather in helping the cell to adapt its functionality to the stress condition.

As a result of all these investigations, researchers not only hope to gain deeper insight into the mechanisms of response to extreme conditions. (In this field, the heat shock response has become a paradigm, while other extreme conditions such as cold, pressure, extreme pH, and high salt concentration are less well studied.) In addition, biotechnologists hope to translate the detailed knowledge of natural mechanisms into agricultural applications. They have already tried to get potato plants to express the winter flounder AFP, hoping to develop a potato variety able to thrive in the high Andes. An ice nucleation protein from *Pseudomonas syringae* already serves in the routine fabrication of artificial snow, and, perhaps, one day, motorists will reach for a biological product to defrost their cars on cold winter mornings.

(1998)

Further Reading
M. Gross *Life on the Edge*, Plenum, 1998.

What Happened Next

Some of the gaps highlighted in this piece have been filled in since it was written. There are now detailed structures of type II antifreeze proteins, which did turn out to be related to lectins. For ice nucleation proteins, there are at least models. Since 2006, AFPs produced in yeast cultures are used in ice cream production, helping to stop the product from growing undesirably large ice crystals.

Colors of the Quantum

In my book, quantum mechanics is definitely cool. Even more so, if quantum mechanical effects, normally restricted to the invisibly small, can be made visible, for instance in a color change. This is what first attracted me to the Q particles, the subject of one of the first five or so science stories I wrote.

If one were to split a red particle into halves, one would expect to get two red particles – normally. For Q particles, however, things aren't quite as simple. These minute grains of semiconductor materials measuring only a few nanometers across can be black, brown, red, or yellow, depending on their size. The "Q" is to be read as a warning that weird quantum mechanical effects have got something to do with this strangeness.

In order to understand these color changes and a few other remarkable properties of Q particles, we should first look at the reason why semiconductors are semiconducting, and why they are different from both conducting and insulating materials. Solid metal and semiconductors are not made up of molecules but rather of vast arrays of atoms. The electrons are no longer found in well-defined clouds (orbitals) surrounding each atom, but are smeared out over the whole solid body in so-called bands. The special thing about semiconductors is that among the two bands, in which the electrons of any solid-state material can reside, the one with the lower energy (the valence band) is fully occupied, while the one with the higher energy (the excited state or conducting band) remains empty (Fig. 18). In this situation, neither of the two bands can transport any charges. They only become conducting when electrons from the lower band are catapulted into the upper one, by some kind of excitation energy, which might come

The Birds, the Bees and the Platypuses. Michael Gross
Copyright © 2008 WILEY-VCH Verlag GmbH & Co. KGaA, Weinheim
ISBN 978-3-527-32287-9

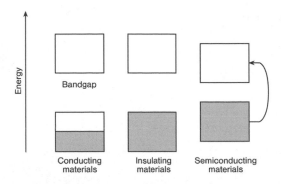

Bandgap

Conducting
materials

Insulating
materials

Semiconducting
materials

Figure 18 How semiconductors work. Electrons in solid materials typically occupy either of two energy "bands," where the lower or valence band may be filled partially or completely. Energy has to be applied to move electrons into the higher or conductance band. In metals, the lower band is only partially filled, so there are free spaces that electrons can easily move to. In insulating materials, the valence band is full, and the conductance band is so high up that electrons cannot normally transfer to it. In semiconductors, however, the energy difference, or band gap, is so small that it can be overcome easily, e.g. by light energy.

from light, heat, or electromagnetic fields. If this excitation is triggered by light waves, certain wavelengths of the light are absorbed in the process, depending on the width of the energy gap between the bands (the "band gap"). This selective absorption of some parts of the continuous spectrum of visible light is the reason why these materials appear to be colored – we see the color of the remaining light which is scattered but not absorbed.

Q particles, then, are half-way between the "infinite" semiconductor material, and a molecule made of only a few atoms. This unique position explains their weird properties. On the nanometer scale, the band gap – which is a characteristic and unchangeable property in bulk semiconductor materials – widens as the particle size shrinks. Accordingly, both the color and the electronic responsiveness of the particles become size-dependent.

How can these particles be made? Making Q particles of just any size is not that difficult. In fact, it can inadvertently happen to chemistry students in their first year. They sometimes have to produce cadmium sulfide – a substance that was used in exposure meters back in those times when these were separate from, and almost as big as, the camera – by bubbling hydrogen sulfide gas through a solution containing cadmium ions. Depending on the exact conditions, they may obtain particles of just a few nanometers diameter, which will readily flow through filter paper. This intriguing effect will usually result in dissatisfaction for the students, who meant to separate cadmium from other dissolved substances in order to get on with their analyses. For the chemical preparation of Q particles, however, this route provides the easiest access to nanosize systems. Chemists exploring the marvels of the nanoworld, such as Horst Weller at the University of Hamburg, Germany, routinely use this approach to make Q particles.

Physicists, in contrast, approach the investigation of quantum mechanical effects in small particles from a different direction. They would profit from the continuing miniaturization of microfabrication methods, defining nanometer sized areas on microchips by either etching away the surroundings, or by applying minute electrical fields. The resulting "quantum dots" can be incorporated into electronic devices.

By careful manipulation with very small voltages, quantum dots can be made to behave like "artificial atoms." If one moves just one electron from the valence band to the conductance band, the former will retain a positive charge, and the latter a negative one. Thus a pair of charges corresponding to the simplest atom (hydrogen) is obtained. Artificial atoms have the advantage that the number of charges can be chosen at liberty, so that researchers can conduct investigations right across the periodic table using only one model system. They are particularly useful for testing the predictions of quantum mechanical theories in simple experiments.

A third field, which is related to both the physical and the chemical approaches to semiconductor nanoparticles, is the chemistry of metal clusters. These chemical entities, which may contain up to a few dozens of metal atoms, also display quantum effects, albeit in an even smaller size range than the semiconductors. One interesting kind of gold clusters has been discussed on page 97. While these three research fields have been developing independently for more than a decade, developments in the 1990s have suggested that they converge on a new and deeper understanding of the quantum effects in nanosize particles.

The combined forces and creativity of all researchers involved will be required to lead these interesting materials to the realm of useful applications. Although they have remained a topic of basic science thus far, numerous applications can be imagined. They could be used, for instance, for the development of better solar panels, involving porous materials coated with Q particles. By varying the particle sizes appropriately, one could adapt the absorbance of the panel to the properties of the sunlight and thus make better use of its energy content.

In chemical reactions, the high proportion of surface exposed atoms in these materials suggests they would be suitable for catalysis. Titanium dioxide nanoparticles, for instance, are being considered for the catalytic treatment of wastewater.

The biggest potential, however, is in the fields of electronics and photoelectronics. Using semiconductor Q particles, one can not only selectively convert light of a chosen wavelength into a current, but one could also make the particles light up by applying a voltage to them. In physical applications, one would certainly wish to exploit the fact that Q particles make it possible to "handle" single electrons or photons. Physicists are already discussing single-electron transistors, and optical switches. These may become elements of tomorrow's supercomputers.

(1993)

Further Reading
H. Weller, *Angew. Chem. Int. Ed.* 1993, **32**, 41.

What Happened Next

While I looked the other way, the term "Q particle" has fallen out of fashion to such an extent that it doesn't show up in Wikipedia, and Google searches list Weller's 1993 review paper as the first hit. I suppose it's been pushed out by an indiscriminate use of "quantum dot," even though I found both concepts useful. Oh well. May the Q particle rest in peace, while quantum dots are still in the race for the first really useful quantum computer, to which I will return below.

Crystals Made to Measure

I used to be very fond of the European Research Conferences organized by the European Science Foundation (ESF), which are small, intimate meetings, typically covering nice interdisciplinary cross-sections of science. In 1996, I attended one such conference, which was on molecular recognition (I was studying the substrate recognition of molecular chaperones at the time). I discovered a really cool paper there, and went on to write my very first News and Views item about this.

Let us time-travel to Paris in 1848. There we find a 25-year-old recently graduated chemist who is seriously puzzled. Why, he keeps wondering, why could tartaric acid rotate the plane of polarized light, while racemic acid – identical by all chemical criteria – would leave it unchanged? He made oversaturated solutions of the optically inactive compound and left it to crystallize overnight on the window-sill of his laboratory. When scrutinizing the crystals with a microscope the next day, he found that there were two types of crystals, which had the same geometry but were mirror-images of each other, like a right and a left glove. Painstakingly using a pair of tweezers while watching through the microscope, he sorted the two types apart, redissolved them separately and checked what the solutions did to polarized light. Those crystals which had the same "handedness" as crystals obtained from tartrate solutions gave a clockwise rotation the same way that tartrate did. And the other solution turned the plane of polarized light by the same angle – but in the opposite direction. Thus, at a time when atoms and molecules were still elusive postulates, our man had established a direct connection between the chirality (handedness) of molecules (as witnessed by the optical activity) and the chirality of the

The Birds, the Bees and the Platypuses. Michael Gross
Copyright © 2008 WILEY-VCH Verlag GmbH & Co. KGaA, Weinheim
ISBN 978-3-527-32287-9

crystals they form. And of course this was only the beginning of the career of Louis Pasteur (1822–1895).

A little less than one and a half centuries later, researchers in the group of Lia Addadi at the Weizmann Institute in Rehovot, Israel, seemed to be following in Pasteur's footsteps in that they were busy sorting crystals formed by mirror-image molecules (enantiomers) using a microscope. This time, the chiral substance was a salt of tartaric acid, namely calcium tartrate. It appeared to be impossible to tell the crystals of the enantiomers apart, as they are symmetrical in appearance, not chiral. If, however, the researchers brought certain types of living cells, such as cultivated kidney cells of the clawed toad *Xenopus laevis* in contact with the mixed crystals, these seemed to be very well able to distinguish between the apparently identical crystals. In the first phase of the experiment they settled exclusively on a certain surface of crystals formed by the RR enantiomer. With a Pasteur-style sorting procedure, measuring the optical activity of solutions derived from the crystals that the cells chose to grow on in comparison to the others, Addadi and coworkers could prove that cell growth on the crystal surface is a reliable criterion for enantiomeric separation of otherwise indistinguishable crystals.

However, the cells that chose the RR crystals in the early hours of the experiment came to experience the downside of their choice after a day or so. The binding of their cell surface molecules to the crystal surface was so tight and rigid that the cells died like insects stuck on fly-paper. On the crystals of the other enantiomer, in contrast, smaller and less tightly bound cell cultures began to thrive on the second day and survived for several more days.

This demonstration of how molecules of the cell surface can distinguish the chirality of the constituent molecules within macroscopically identical crystals is just one example of the manifold and often amazingly specific interactions between biological macromolecules and small molecule crystals. In 1994, Addadi and her co-workers showed that antibodies that they had raised against different salts of uric acid and against the neutral analogue allopurinol can specifically promote the formation of the type of crystal they had been raised against. Although the specificity of the antibodies was selected for the recognition of fully grown crystals, they seemed to stabilize the very first assemblies of say 20 to 30 molecules during the nucleation of crystal growth. In a similar way as those antibodies which researchers have raised against transition state analogues of simple chemical reactions, and which were found to act as catalysts by lowering the energy of the transition state as enzymes do, Addadi's antibodies seem to be specific catalysts for the more complex assembly reaction that leads to crystal formation.

More recent work from her group has resulted in monoclonal antibody preparations, which consist of just one specific protein sequence each, so that structural investigations become possible. Structure predictions for one particular antibody against cholesterol crystals revealed an intriguingly sharp kink in the recognition site, which exactly matches the edges found on the surface of the cholesterol crystal.

Thus far, one might be forgiven for thinking that such antibodies are just fascinating biochemical toys. But in fact, they may turn out to be models for a physiological process of immense medical relevance. The symptoms experienced in a fit of the joint disease gout are thought to arise from crystals of a uric acid salt accumulating in a joint, which are then recognized by the antibodies of the immune system as intruders (even though the same compound in soluble form does not trigger an immune response), which in turn leads to an inflammatory response in the joint. The specificity results from Addadi's group suggest that a similar recognition in the affected joint is the most likely cause of the inflammation. They also invite the speculation that the immune system may be making things worse by supplying large numbers of antibodies against the offensive crystals, which unintentionally catalyze the formation of more crystals. This might explain why the threshold for fits of gout is lowered as the disease progresses – a case of immunity gone wrong. Moreover, the immune response cannot make the offending crystals disappear, as its cleaning up systems are optimized to be efficient against cells, viruses, and biomolecules, rather than against small molecule crystals.

However, interactions between proteins and crystal surfaces can also be immensely useful for organisms that depend on controlling crystal growth, as is the case in biomineralization and in protection from freezing (see Life on the Rocks, page 161). Finally, there is a further group of extremely important natural processes relying on the interactions between biological macromolecules and the crystalline or amorphous phases of solid substances. It includes the processes that form our bones and teeth, and those that provide invertebrates such as mollusks and snails with a rigid outer shell, covering the formation of more than 60 different minerals and a wide range of phenomena, which are collectively called biomineralization.

The biomolecules controlling these processes can decide whether deposition of the mineral from the solution leads to an amorphous or to a (micro)crystalline phase, they can trigger crystallization, shift a bias between different crystal forms (morphologies), guide or limit the crystal growth directions in space, and stop the process altogether. In most cases (but not in our bones) the controlling agents belong to a class of proteins that are so exotic that they are often referred to as "unusually acidic macromolecules." (The first example was a protein retrieved from teeth during the 1960s, which was found to contain 40 mol percent of aspartic acid.) Generally, these molecules have aspartic acid in every third or every other position, and they also contain exceedingly high percentages of phosphorylated amino acids, mainly phosphoserine. All these unusual properties lead to the problem that these proteins are extremely difficult to analyze and characterize with conventional methods. Often even determining the molecular weight causes problems. Presumably they can form extended beta sheet structures similar to those of ice nucleation proteins. Although the control of mineralization by these molecules can be mimicked in the test tube, their mechanisms are still controversial. It would be conceivable that the acidic proteins act at membranes and/or in free solution by influencing the formation and orientation of nuclei and/or the growth of the crystals.

In many cases – just think of the spirals of snails' shells always wound up in the same direction – the control of crystal growth by chiral biomolecules leads to macroscopic objects which have chirality just as Pasteur's crystals did. Intriguingly, this may even happen if the molecular building blocks (calcium phosphate in the case of the snail) are achiral.

(1995)

Further Reading

M. Gross, *Travels to the Nanoworld*, Perseus, 1999.

What Happened Next

The interaction between hard crystals and presumably soft proteins has continued to fascinate me in various contexts, so I have covered several other advances in this field, particularly the molecules involved in the formation of diatom shells, see page 218. Addadi's student Joanna Aizenberg has been very successful at the biominerals/nanotechnology interface, creating a splash with the 2003 discovery of "fiber optics" in a marine sponge.

The Incredible Nanoplotter

Nanotechnology is pretty cool, I hope. (Though I haven't
checked with the kids yet.) After publishing *Travels to the
Nanoworld*, where I explored the ways in which nanotechnology
can learn from biological nanosystems, I tried to keep up with
this fast-moving field by writing the occasional piece on new
nano-tools and gadgets. Dip-pen lithography struck me as one
of the coolest. Plus, as a writer, I am bound to be interested in
new writing technologies. This is also an early (and over-length)
example of my favorite writing format, the 800-word minifea-
ture, which saw its golden era in 2000–2003, when I regularly
contributed one page "frontiers" to *Chemistry in Britain*.

In December 1959, the famous physicist Richard Feynman challenged his colleagues to shrink a book page 25,000-fold. This factor
brings the height of a large format paperback (21 cm) down to less
than a hundredth of a millimeter (10 microns) so it would roughly fit
on a red blood cell. The height of a typical capital letter (Times Roman,
12 pt) would be reduced from three millimeters to a tenth of a micron,
thus it could be hidden from sight by a single bacterium. Feynman's
after-dinner speech encouraging scientists to "think small" is now
seen as the birth hour of the futuristic concept of nanotechnology, the
technology based on machinery that is as small as the molecules
found in living cells.

Twenty-six years went by before Feynman's vision became reality.
In 1985, Stanford graduate student Thomas Newman programmed an
electron beam apparatus to copy the first page of Dickens' *A Tale of
Two Cities* by etching each of its letters into a surface smaller than a
needle tip. The letters were drawn as simply as possible, and the lines

The Birds, the Bees and the Platypuses. Michael Gross
Copyright © 2008 WILEY-VCH Verlag GmbH & Co. KGaA, Weinheim
ISBN 978-3-527-32287-9

and dots measured around 60 atoms across. Electron beam lithography is still the most subtle instrument we have to create very small structures, but the fact that features must be drawn sequentially (one dot after another) rather than printed all at once means it is not commercially viable for large copy numbers.

In manufacturing, light waves are used to print small patterns in large copy numbers, copying from a single original (mask) often made by an electron beam. This is the way computer chips are made. Similar methods are also used in the production of micro-electro-mechanical devices (MEMS), including the sensor that triggers most of the car airbags in use today. But, as the wavelength of visible light is around half a micron, it is very difficult to extend these methods to the 0.1 micron (100 nanometer) scale envisaged by Feynman. As the wavelengths required are shifted deep into the range of the invisible ultraviolet (UV) region and eventually into the X-ray range, photolithography becomes more and more tricky. The lack of suitable optics to focus X-rays limits this technique severely.

An electron beam, in contrast, is easily focused even at very small scales. Its use is only restricted by the quantum mechanical effects that make life interesting on the atomic scale. It can produce features as small as ten nanometers across. Typically, it is used for making prototypes or to create structures needed in research, as well as for the masks used in photolithography. But trying to scale up this handwriting method to the production of large copy numbers with reasonable efficiency would be like getting monks writing with quills to compete with a printing press.

The introduction of a new kind of nanometer scale "dip-pen," however, has started to shift the balance in favor of the monks. In 1999, Chad Mirkin at the Northwestern University in Evanston, Illinois, used the atomic force microscope (AFM, an instrument normally used for imaging invisibly small surface structures) to apply a delicate handwriting of molecular "ink" to a sheet of "paper" made of gold. They made use of an effect that often bothers people who want to use AFM for its original purpose. The very small distance between the surface in question and the extremity of the AFM tip attracts humidity by acting as a capillary. Unless the experiment is run in an absolutely dry environment, there will always be a water meniscus linking the tip to the substrate. In the new technique, which Mirkin called "dip-pen nanolithography" or DPN, this water meniscus serves as a

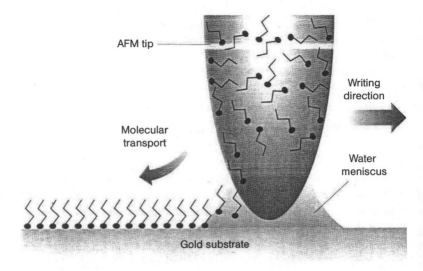

AFM tip

Writing direction

Molecular transport

Water meniscus

Gold substrate

Figure 19 Dip-pen nanolithography (DPN). This method, developed by the group led by Chad Mirkin, relies on the transfer of a liquid from an AFM tip to a surface.

slide on which the organic molecules that make up the ink slide down onto the gold surface.

Feeding the ink to the tip, the researchers realized that the speed with which it spread onto the gold surface depended critically on the dimensions of the water meniscus, which in turn could be controlled by varying the humidity of the surrounding air. This allowed them to draw dots and lines of very exactly controlled thickness. The thinnest lines they can produce are 30 nanometers wide, which would be just right for the 25,000-fold reduced print page required by Feynman's challenge. What stopped them from going even further was not the dip-pen technique itself, but the grain structure of the gold surface they wrote on.

As with Newman's miniaturized page from Dickens, this achievement was quite impressive as a one-off, but not suitable for technical applications, where you would nearly always want to build large numbers of identical copies of a given structure. Therefore, Mirkin's lab went on to explore the possibilities of building a machine with multiple nano-pens writing exactly the same pattern. In June 2000, they reported the first success of this endeavor in *Science* magazine.

They realized to their own surprise that the speed with which the ink runs from the tip onto the substrate does not depend on the force with which the two are pressed together. This allowed them to arrange several AFM tips in a way that they can all be controlled by the same system. Subtle differences between the vertical forces applied to each of the tips, as could result from irregularities of the surface of the writing paper, could safely be ignored. Thereby the group created what they call a nanoplotter, working both serially and in parallel. The tips can pick up their ink from small pieces of filter paper. Similarly, they can be "rinsed" by touching a filter paper soaked with a suitable solvent.

Having demonstrated the principle on a nanoplotter with two tips, they proceeded to an arrangement including eight tips, of which again only one was controlled by the operators, while the other seven tips were blindly copying whatever the first one did. Drawing simple geometrical figures such as squares and octagons in eight copies, they found that the measurements of the copies were identical within satisfactory limits. They even claim that "there is no reason why the number of pens cannot be increased to hundreds or even a thousand" without additional control mechanisms.

Thus, dip-pen nanolithography will no longer be restricted to manual production of single copies. Moderate print runs of the Dickens page could be created within a day. More importantly, intelligent choice of ink and paper will allow nanoscale engineers of the future to create nanometer scale elements that today's chip manufacturers can only dream of. Chemical nanofactories, where small numbers of molecules are guided to meet each other in a chosen order of reactions could be built. Electronic elements including computer chips could be so small that they could fit into the dot on this "i," or even into a living cell. The nanotechnology revolution that has been speculated on since the late 1980s has just come a few microns closer.

(2000)

Further Reading

R.D. Piner et al., Science, 1999, 283, 661.

What Happened Next

Mirkin's group has continued to explore the possibilities of DPN with great success, creating protein nanoarrays and nanostructures with this technology.

Further Reading

K.-B. Lee *et al.*, *Science*, 2002, **295**, 1702.

Silencing the Cacophony

Continuing the theme of "cool tools," the gene silencer described below must be the most elegant tool of modern molecular biology. I caught this one quite early, when the mechanism was still far from understood and the name RNAi was still unfamiliar, even to scientists. It soon became big news, culminating in the award of the 2006 Nobel Prize.

Imagine you are at a gigantic concert. An orchestra of 25,000 musicians plays something that is supposed to be music, but the multitude of sounds and rhythms adds up to a cacophony that makes your head ache. You fail to make out any melodies, harmonies, or motifs. There is no conductor in sight. Various instruments or groups of instruments appear to influence each other following mysterious rules. Thousands of brass players cover up the soft-voiced instruments, but only if they aren't drowned themselves by dozens of organs.

While you are desperately plugging your ears and planning an escape route, your seat neighbor seems to be having the time of her life, listening to the music in transfixed ecstasy. In her hand she holds what looks like a remote control of a TV set or DVD player. Noticing your bewildered glance, she hands you the control, smiling encouragingly. You press a button at random, and the miracle happens. Suddenly, you hear only the string section of the mega-orchestra, playing in perfect harmony. Further buttons enable you to add or eliminate instruments and groups of instruments, enabling you to appreciate the role of each individually or in context with a few others. Thus, you gradually get an impression of how the apparent chaos of the whole is made up of meaningful connections between the contributors.

The Birds, the Bees and the Platypuses. Michael Gross
Copyright © 2008 WILEY-VCH Verlag GmbH & Co. KGaA, Weinheim
ISBN 978-3-527-32287-9

For researchers trying to pick apart the cacophony of our 25,000 or so genes in the human genome, such a magical remote control would be the ultimate gadget. Only by switching genes on and off will they be able to figure out how they harmonize with each other.

This gadget may not remain a dream forever. In a first step towards its realization, researchers have developed – or rather: nicked from nature – a silencer that can mute individual instruments of the genomic orchestra. While silencing individual genes may seem like a small step, it can already be a great help in understanding the genome sequence, which provides a list of the genes, but without details on what role they are going to play in the whole.

Chopping up Suspicious RNA

Before the information contained in a gene can materialize as a protein, the gene has to be transcribed into messenger RNA. Unlike the famous double-helix structure of DNA, RNA typically comes as a single strand. (Hairpin-like structures where the strand forms a double helix with itself are common in structural RNA, but highly undesirable in messenger RNA, as they would block its function.) Double-stranded soluble RNA was not known to occur in the repertoire of the living cell.

In some groups of viruses, however, double-stranded RNA is found as the genetic material. Therefore, many organisms, ranging from plants to fruit flies, have a defense mechanism that is triggered specifically by RNA double helices. If the cell police spots a suspicious double-stranded RNA, it will chop it up into short fragments of around 21 base pairs. Being a rather drastic kind of police force from the zero-tolerance school, it will not only destroy these fragments without a trace. It will also remove all messenger RNAs that happen to carry matching sequences, based on the assumption that these messengers might have originated from the virus.

In 2000, Emily Bernstein and her coworkers at Cold Spring Harbor Laboratory (New York) identified the enzyme that recognizes the foreign RNA and chops it up in the fruit fly *Drosophila*. Borrowing the name of a kitchen tool, they called the enzyme "dicer." Around that time, the defense mechanism, which had previously been known by different names in different groups of organisms, came to be known by the name RNA interference, or RNAi.

At that point, it was still unclear whether mammals, too, use this defense mechanism to fight off RNA viruses. In 2001, Thomas Tuschl and colleagues at the Max Planck Institute at Göttingen, Germany, answered this question by applying the insights obtained from *Drosophila* to mammalian cell culture. At the same time, they invented a gene silencer.

In *Drosophila*, the enzyme chops double-stranded RNA into fragments containing exactly 21 base pairs. In order to check whether the analogous mechanism exists in mammals, Tuschl's group synthesized such fragments in the laboratory and chose their sequences such that they matched a reporter gene which they had inserted into the genome of their cell cultures, such as the gene for the luminescence enzyme from fireflies, luciferase. In the presence of the substrate, luciferin, mammalian cells with this gene will glow like fireflies. However, when the researchers injected their matching fragments of 21 RNA base pairs into these cells, the luminescence was quenched measurably. Thus they had shown two important things in one experiment. First, mammalian cells have an analogous system that recognizes double-stranded RNA and suppresses any homologous messengers. Second, the system can be used to silence other genes within the cell in addition to viral ones.

Strictly speaking though, the luciferase reporter gene was a foreign intruder, much like a virus. Therefore, Tuschl's group conducted a further experiment to show that even the cell's very own genes can be silenced this way. For this, the researchers used the cell line derived from the tumor of a cancer patient, known as HeLa cells. In this cell culture, they successfully suppressed the genes for two proteins that are normally found in the membrane of the nucleus.

Switching off Genes at Will

It appears that the mechanism functions in a very similar way in mammals and insects. One crucial criterion for its use in silencing is that the artificial RNA fragments need to be shorter than 30 base pairs, because longer fragments will trigger other defense systems in mammals.

Thus, the RNAi method allows researchers to switch off any one of the around 25,000 genes in the human genome at will. As genes that are related to each other and form gene families will typically have some parts of their base sequences in common, it is also possible to silence groups of genes. In contrast to the more primitive method of adding "antisense" DNA or RNA, which is a longer piece of nucleic acid that simply blocks the messenger RNA, the new method requires less than a thousandth of the material.

Thus, the usefulness of the new gene silencing tool has become immediately obvious, even before the details were known of how the cell responds to the presence of the RNA. Like the imaginary concert experience mentioned above, our understanding of the genome is much improved by the help of a silencer.

(2001)

Further Reading

S.M. Elbashir et al., Nature, 2001, 411, 494.

What Happened Next

RNAi established itself immediately as an essential tool in genomic research. Further research into the mechanisms revealed, however, that Nature had already invented the silencer function long before humans re-invented it. Parts of our genome code specifically for regulatory RNA fragments, so-called siRNA (small inhibitory RNA), which trigger the RNAi response and thus form an important part of the network by which the genome regulates itself.

In 2006, Andrew Fire at the University of Massachusetts and Craig Mello of Stanford received the Nobel Prize for physiology or medicine for the discovery of RNAi.

From E-ink to E-paper

Those big old-fashioned TV screens and computer monitors were anything but cool and are now rapidly losing ground to flat screens. But could electronic displays become as flat and flexible as a sheet of paper? Now that would be cool …

Please spare a few seconds to consider the sheet of paper these words are printed on. It's such a versatile and useful display type. It's thin, light, flexible, and reasonably robust. You can read it anywhere, you can highlight certain words or cross them out and scribble your own comments instead, and you could pull out the page to file the text, or to roll it into a ball and try hitting the waste basket across the room.

There is just one thing wrong with it – once you're finished with all of the above uses, it just stays where you left it and remains basically a waste of space and resources. Even if you eventually put it in the re-cycling container, its rebirth as a new sheet of paper will require en-ergy. Just think how much easier and greener it would be if you could just press your pen on a little icon at the bottom of the page, and the same physical page would then show you the next page of the book, or some other text. Or imagine when you've just finished the first vol-ume of Harry Potter, you speak the magic words, open the book at the front again and you can read the second volume on the same pages.

Some recent advances suggest these fantasies might soon become reality. Back in 1998, the group led by Joseph Jacobson at MIT pre-sented an "electrophoretic ink," which can be printed onto any mate-rial and can be switched between black and white appearance by a small voltage. E-ink, which is now commercialized by a company of the same name (www.eink.com) is a liquid containing microcapsules. Each capsule encloses charged white pigments in a black liquid.

When an electric field is applied, the white particles will migrate towards the positive charge. They will stay on that side until a field of the opposite polarity is applied. Unlike an LCD display, this ink does not require any energy to carry on displaying a black or a white pixel.

Now the group of John Rogers at Bell Laboratories, in collaboration with the E Ink company, has taken the next step. Combining the very latest techniques in microfabrication, exotic semiconductors, and e-ink, they built what you could call the first electronic piece of paper: a display that is only 1 mm thick, doesn't mind being bent, and has the look of paper.

The principle that allows the reader to turn the page electronically is a simple array of transistors arranged on an x/y grid. There has to be one transistor for each pixel, and the prototype e-paper has 256 of them: 16 rows with 16 pixels. Each pixel is defined by a gold electrode connected to the drain of its transistor and can be addressed individually by a small voltage applied to the transistor via the appropriate contacts corresponding to its x/y coordinates. Switching the whole display takes about one second.

A key technology for the production of robust transistors on a flexible substrate was the use of microcontact printing (μCP), a "soft lithography" technology pioneered by George Whitesides' group at Harvard. This involves a microfabricated "rubber stamp" typically made of polydimethylsiloxane (PDMS), which can transfer a patterned monolayer of organic molecules onto a substrate. In this case, the substrate is a gold layer (20 nm) which serves as the source/drain level of the transistors. Printing the desired pattern onto the gold allows it to be protected in the subsequent etching step, whereby the appropriate features (source and drain contacts and pixel electrodes) are produced.

So when you're looking at a sheet of electronic paper, you're dealing with a little bit more chemistry than you would find in the ordinary wood pulp product. From top to bottom, there is first a transparent layer of indium tin oxide serving as an electrode shared by all pixels. Application of a positive charge to this electrode turns the entire display white except for those pixels that have a greater charger on their gold electrodes. Then comes the layer with the famous e-ink microcapsules, followed by the transistor architecture: the gold electrodes, an organosilsesquioxane spin-on glass serving as the dielectric, and the gate electrodes made of indium tin oxide. Finally, at the bottom of the whole sheet, there is a rather ordinary drinks bottle material, polyethylene terephthalate (PET) serving as a flexible support.

Although the current thickness is more like cardboard than paper, the low reflectivity and sharp contrast of the display is meant to give the reader a very papery impression. Just remember not to throw it in the recycling bin.

(2002)

Further Reading

B. Comiskey *et al.*, *Nature*, 1998, **394**, 253.
Y. Xia and G. M. Whitesides, *Angew. Chem. Int. Ed.*, 1998, **37**, 551.
J. A. Rogers *et al.*, *Proc. Natl Acad. Sci. USA*, 2001, **98**, 4835.

What Happened Next

Electronic paper, like e-books, is one of the things that always seem to be around the next corner, and several companies appear to be competing to overcome the last obstacles and launch a commercially viable product. As of November 2007, however, all the books and magazines in my household are on old-fashioned wood pulp (or recycling) paper.

Spinning Lessons

They may have eight hairy legs, a very small brain, and a big image problem, but there is at least one thing that spiders can do much better than we can, namely produce extremely strong fibers.

Nature is in many ways a better engineer than humankind. Whether you look at the ways that diatoms, mussels, or snails make their shells, butterflies seem to change their colors, or trees withstand strong winds, there is always a lesson for human engineers to learn. And sometimes, nature's engineering is so clever we still have trouble figuring out how it works. But when we do, the rewards to be gained from the lesson can be enormous.

One of the most spectacular examples of how nature beats our best efforts is spider silk. Like our hair, sheep wool, and silk clothes, it consists mostly of protein. But the polypeptide chains are aligned and interwoven in mysterious ways that make the product much stronger than these materials. Optimized by evolution to be able to stop an insect in full flight, this is in fact the strongest material we know in strength per weight terms. If you compare a spider's thread with a steel wire of the same diameter, they will be able to support roughly the same weight. But the silk is six times lighter, so it is really six times stronger than steel, and the spider wins every time.

So why are suspension bridges still dangling on steel ropes rather than silken ones? The trouble is we can't make spider silk as nicely as the spider can. Sure, we can express the proteins of which it is made in other organisms, including goats that will have spidroin in their milk (to which I will come back later), and we will soon be able to spin those into some kind of fiber, but given our limited understanding of

 The Birds, the Bees and the Platypuses. Michael Gross
Copyright © 2008 WILEY-VCH Verlag GmbH & Co. KGaA, Weinheim
ISBN 978-3-527-32287-9

the processes going on in the spider's silk gland, the result may not live up to the natural product.

A small scattering of biologists in various laboratories around the world are trying to get behind the spiders' secret. First they need the right genes. Until recently, only very few of the DNA sequences of silk protein genes were known. In 2001, John Gatesy and Cheryl Hayashi with their coworkers at the University of Wyoming in Laramie presented a comprehensive overview of gene sequences from a wide variety of eight-legged silk producers, including tarantulas and other animals that separated from the "true spiders" more than 200 million years ago. They showed that the amino acid sequences are extremely diverse between species. About the only thing they all have in common is the occurrence of unusual repetitive sequences following four simple patterns: polyalanine (An), alternation of glycine and alanine (GA), and combinations of glycine with a small subset of amino acids (X) with or without proline: GGX and GPGGX.

Given that these motifs have been retained (or evolved convergently) over a time span of more than 200 million years, it appears likely that their properties hold important clues to how the silk proteins interact to form silk. So far, the only structural information we have is about the finished silk, where it is known that the alanine-rich repeats occur in quasi crystalline domains, while the glycine-rich repeats adopt more disorderly states that are poorly understood.

From these genes, the spider makes the corresponding proteins, called fibroins. Nothing special there – proteins can make hair and wool and the brittle type of silk that insects make their cocoons of. But these materials aren't tough enough for spiders. To make the special, stronger-than-steel brand of protein filament, spiders have a special silk gland, a complex structure where the magical transformation of protein solution into silk thread takes place. Although this transformation is only poorly understood so far, it is known that it involves a substantial increase in the proportion of the protein chain that is arranged as a beta-pleated sheet.

In a 2001 review in *Nature*, Oxford zoologists Fritz Vollrath and David Knight, who have been studying silk production in the orb spider *Nephila clavipes* for many years, summed up the current knowledge. First, it should be noted that these spiders do not have a single kind of silk gland, but seven pairs producing seven different types of silk for different uses, and even the protein composition in these glands is significantly different. The one that has been best character-ized is the dragline silk which is produced by the major ampullate gland.

This gland consists of three major regions: a central bag (B zone) flanked by a tail (A zone) and the duct (D) leading towards the exit (Figure 20). The lining of the A and B zones contains the cells that se-crete the protein material, the major component of which is a 275-kDa protein containing the polypeptides spidroin I and spidroin II. The A zone specializes in the spidroin protein, which forms the strong core of the thread, while the B zone is believed to secrete the as yet poorly understood glycoprotein material that ends up coating it. To be se-creted from a cell, proteins must be wrapped up in membrane bub-bles called secretory vesicles. In the A zone cells, these vesicles con-tain protein filaments, the precise structural organization of which is still under investigation. The B zone vesicles contain liquid crystals of the coat glycoprotein. Knight and Vollrath think that the liquid crys-talline state has an important role to play in the production of the silk thread, to which we shall return later.

Let us follow the route of a spidroin molecule from secretion through to the finished thread. On leaving the cells of the A zone (as the secretory vesicles merge with the cell membrane and empty their contents to the outside), our protein finds itself in a small spherical droplet with lots of other spidroin molecules. The protein concentra-tion in the whole gland is around 50% – higher than in most protein crystals. Most proteins would aggregate into insoluble lumps at much lower concentrations. This highly viscous protein mass flows down the A tail into the bag (B zone), where it gets coated by the secretion of the B zone cells. At the exit of the bag the liquid is funneled into the much narrower duct (D). During this transition, the droplets are slow-ly distorted into long thin shapes aligned with the direction of the flow. It is assumed that a similar transformation happens to the molecules. Initially they must have been in a rather compact conformation to avoid aggregation, but as they move into the duct, they are stretched out and aligned in a way that will eventually allow them to form those intermolecular links that hold the thread together.

The spinning solution or dope is now in a liquid crystalline state, with proteins aligned in an orderly fashion, but still able to slide past each other. This is thought to be an important part of the spider's se-cret weapon. As the material is slowly and gently narrowed down in

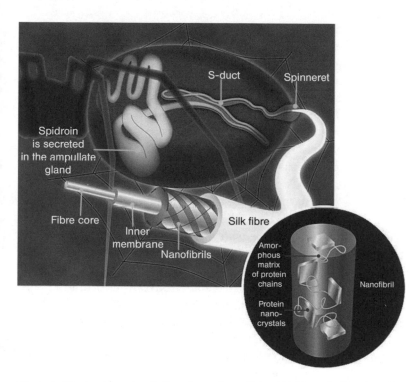

Figure 20 How a spider spins its thread. The main protein component, spidroin, is secreted in the ampullate gland, B. Its material properties change in significant ways during its passage through the S-duct.

(From: Richard D. Pines, Jin Zhu, Feng Xu, Seunghun Hong, Chad A. Mirkin, Science 283, 661 (1999). Reprinted with permission from AAAS.)

the first two legs of the tripartite duct, molecules have time to orient themselves in a favorable way so they can eventually form intermolecular beta sheet interactions and possibly disulfide bonds when it comes to making the actual thread. This step happens at a point ca. 4 mm before the exit, and it happens quite suddenly. Although the molecular details are far from clear, it is thought that as the dope is drawn out to a thin thread that separates from the walls of the duct, the molecules align even further and form hydrogen bonds defining the complex beta sheet patterns found in the finished product. In the process, the protein becomes more hydrophobic and ejects some of the water content it has been carrying until this point. Finally, most of the water is stripped off the surface when the thread leaves the exit spigot, helping the spider to avoid water loss and making its thread even tougher.

This broad picture drawn by Vollrath and Knight combines anatomical with some structural data. However, the exact details of the crucial structural transitions are far from understood. The trouble is that the most powerful tools for the determination of protein structures, X-ray crystallography and NMR, require a protein crystal or a homogeneous solution, respectively. As yet, there is no method that could give you the atomic detail structure of a protein molecule flowing down the duct of a spider's silk gland.

And yet, even in the absence of a full understanding in molecular terms, maybe one could copy the spider's technique on a microscopic scale, by supplying a dope with the right protein composition and passing it through a spinning device modeled on the spider's gland? The only comparable synthetic material, aramid (KEVLAR™, the fiber used in bullet-proof vests), is spun from hot sulfuric acid. Thus, an ambient temperature process leading to something similar would be very attractive, even if the resulting fiber turned out to be only as good as Kevlar, and not quite as good as authentic spider silk.

But first, you need to produce the proteins in reasonable quantities. Unlike silk moths, spiders have an aggressive territorial behavior, which means they won't be cooperating with any ideas of high-throughput farming. Expressing the silk proteins in bacteria or yeasts doesn't work either. The curious repetitive nature of their sequences invites the microbes to take shortcuts and produce abridged versions of the protein chains.

Thus, if you want to use the silk to catch fighter jets rather than flies, you'd better get an animal that can produce more than a few milligrams of the precious material. The company Nexia Biotechnologies Inc. in Montreal, Canada, was the first to succeed in breeding goats that are genetically modified in such a way that they secrete spidroin protein in their milk. It turned out that the secretory cells of mammary glands aren't that different from those of silk glands, only there are a lot more of them in a goat, which makes milking goats a lot more economical than milking spiders.

Since the summer of 2000, Nexia boasts the possession of two African dwarf goats, Peter and Webster, who have been shown to carry the appropriate spider gene. A couple of breeding generations later, there will be a flock of females producing spidroin in their milk by the gram. Nexia keep mum about which way exactly they want to spin that milk-silk protein into strong fibers on an industrial scale. As soon as they can do that, however, applications ranging from surgical threads through to missile protection and aviation security will be conquered rapidly by the new material.

Although some of the applications envisaged are substantially scaled up in comparison to a spider's web, there is also a case of scaling down from there. In an attempt to turn a visible thread into an invisibly thin nanowire, the group led by Michael Stuke at the Max Planck Institute for Biophysical Chemistry in Göttingen stripped spider silk down to the core, using UV laser technology. They obtained very strong nanowires, currently as small as 100 nm diameter. Plans for the future include coating this thread in metal to make it conductive.

But even when we can copy the spider's thread and use it on various length scales, the hairy little arthropods can still go one better. As Stefan Schulz and his coworkers at the Technical University of Braunschweig, Germany, reported in 2000, the female tropical spider *Cupiennius salei* leaves a thread marked with sex pheromones, which induce any male of her species to vibrate excitedly. The vibrations are transmitted through the thread, which rapidly switches from a role of odor dispenser to that of a phone line. The female vibrates back, and you can figure out the rest for yourself. I wonder whether anybody wants to set up a company banking on that technology ...

(2001)

Further Reading

M. Papke *et al.*, *Angew. Chem. Int. Ed.* 2000, **39**, 4339.
J.Gatesy et al. *Science*, 2001, **291**, 2603.
F.Vollrath and D.P.Knight, *Nature*, 2001, **410**, 541.

What Happened Next

In November 2007, I interviewed Fritz Vollrath as part of my research for a feature article for Oxford Today. I found out that some of the hopes mentioned in the earlier piece have failed to materialize. In particular, the attempt to produce artificial spider silk from the spider proteins expressed in goat milk has failed. Vollrath and his coworkers have developed methods to analyse the silk dope and to predict whether it will yield proper silk, but the only creature that can produce spider silk remains the spider. Current hope rests on genetic manipulation of more "farmable" animals such as moths, in order to get them to produce spider silk in large quantities. Japanese researchers have already announced that they have tricked silkworm caterpillars into spinning cocoons that contain 10 % spider silk.

In terms of application prospects, Vollrath has high hopes for biomedical fields including joint repair. As some spiders can produce half a dozen different kinds of silk with specific properties for specific tasks, scientists should one day be able to adapt synthetic materials derived from spider silk to the precise requirements of surgical applications.

Traffic-light Proteins

The use of Green Fluorescent Protein (GFP) as a gene marker in living cells must count as one of the coolest inventions in the last two decades. Plus, it happened at the beginning of my writing career, so I covered the original work and then sat back and watched it grow to one of the biggest things in molecular biology ever (see page 101). A few years later, the palette expanded to include different colors, which is the topic of the following story:

Applications of the green fluorescent protein (GFP) from the jellyfish *Aequorea victoria* have spread epidemically ever since the 1994 paper which demonstrated that GFP can be used as a marker in gene expression. The GFP gene, now available as a commercial kit, can be coupled with the gene of interest. If and only if GFP is produced, the cells will shine green under a UV lamp. It really is that simple, no other chemicals required, no strings attached. Biophysicists have also shown a lot of interest in the small part of the protein that sends out the light (the chromophore) because of its highly unusual structure which results from modifications the protein performs on itself.

In 2000, researchers cloned and characterized an equally intriguing cousin of GFP called DsRed. This is a related protein from corals (genus: *Discosoma*), which fluoresces red and is partially responsible for the characteristic pinkish hue of the coral. Detailed studies reported by the laboratory of Roger Tsien at the University of California at San Diego in 2000 have shown up some advantages and disadvantages of this system in comparison with GFP.

Although fine-tuning GFP can result in a range of colors, none of the GFP variants reaches the wavelength range of red light. Hence the 583-nm emission of DsRed is a very welcome discovery for anyone who wants to be able to use multiple markers and detect their emissions simultaneously. Exchange of a single amino acid residue can even shift it up to 602 nm. Like GFP, DsRed forms its chromophore autocatalytically by fusing neighboring amino acid into an imidazole ring. It first forms a structure similar to the one in GFP and then extends it to achieve the characteristic long-wavelength emission.

The latter reaction is extremely slow, which is a drawback for some applications. But positive thinking can turn this into an advantage, as Alexey Terskikh and his coworkers at Stanford University have demonstrated. They deliberately created a slowly changing mutant of a similar red fluorescent protein and showed that it can be useful as a "timer" in developmental biology. In these experiments, green light reveals cells where the gene of interest has recently been activated, a yellow to orange color shows continuous activity, while red fluorescence indicates the gene has been switched off.

Another potential disadvantage of DsRed is that it likes to associate into tetramers and maybe even into higher oligomers. Given the amount of experience that researchers have accumulated with the homologous GFP, however, it is safe to predict that quite soon there will be genetically altered DsRed variants that are monomeric, mature faster, and shine in every imaginable hue of red. And watch out for the molecular traffic light that switches from green to red at the push of a button.

(2001)

Further Reading

L.A. Gross *et al.*, *Proc. Natl Acad. Sci. USA*, 2000, **97**, 11990.

What Happened Next

Fluorescent markers in the style of GFP and DsRed are big business in the laboratory supplies market now, with dozens of variants fighting for the attention of molecular biologists. I haven't followed this development all that closely, but if there is any use for molecular traffic lights, I am sure somebody is selling them already.

A Cool Receptor

Now this story is cool in every conceivable sense of the word: the discovery of a receptor molecule responding to both menthol and low temperatures. Cooooool …

Science knows surprisingly little about how our senses (apart from the well-characterized eyesight) work on a molecular scale. In particular, molecules responsible for taste and temperature sensation have only become accessible to research since the mid-1990s, when the combination of molecular cloning techniques and genome-wide database searches gave researchers a better chance to track down these rare and elusive membrane proteins. It emerged that there is an interesting link between the "hot" sensation caused by chili peppers and that caused by high temperatures: both are mediated by the same membrane receptor, which the group led by David Julius at the University of California at San Francisco cloned in 1997 and named VR1 (for vanilloid receptor type 1, as the chili pepper's active ingredient, capsaicin, is from the family of the vanilloids).

Subsequently, Julius' group has applied the same logic that helped to identify VR1 as a receptor for noxious heat to the sensation of cool. They chose a chemical that elicits a cool sensation, namely menthol. Then they identified a group of nerve cells that respond to this chemical and to cold temperatures. Rat trigeminal cultures seemed to perform better than the dorsal root ganglion cells often used in such research. Using these cells, they characterized the physiological properties of the as yet putative receptor, in order to find out how it can best be caught. The response to the cool sensation appeared to be mediated by an outward flux of cations, including – most importantly – calcium ions.

The Birds, the Bees and the Platypuses. Michael Gross
Copyright © 2008 WILEY-VCH Verlag GmbH & Co. KGaA, Weinheim
ISBN 978-3-527-32287-9

Therefore the researchers screened the complementary DNA (cDNA, i.e. DNA derived from the messenger RNA found in the tissue and thus representing the genes that are active) by using calcium imaging techniques to check whether cell cultures expressing one of the cDNAs did or did not respond to menthol. They obtained a single cDNA which conferred to its host cells an ion flux response to cold, to menthol, and also to the cooling compounds icilin and eucalyptol. The combined results of further investigations in different types of cell cultures showed that the encoded membrane protein has all the properties required to explain the physiological response of the cold-sensitive neurons from which the study started. It was therefore called the cold-menthol receptor type 1, short CMR1.

CMR1 turned out to be a membrane protein of more than 1100 amino acids, with marked sequence similarities to a group of ion channels of thus far unknown physiological role, the so-called long TRP channels. Unlike most of these, however, CMR1 does not contain an enzyme domain dangling off the cytoplasmic side of the membrane. Searching the human genome for cousins of the new receptor, the researchers found a gene called trp-p8, which was believed to be specifically expressed by healthy prostate epithelia and a variety of tumors. As yet it is not known whether this gene is also expressed in human cold-sensitive neurons and thus qualifies as the equivalent of rat CMR1.

At the same time, another research group arrived at the same conclusions coming from a different direction. Peier and colleagues had investigated TRP channels with a view to clarifying their physiological functions. As both of the known heat sensors VR1 and VRL-1 belong to this family, too, cold sensation was a reasonable hypothesis to check. This group identified the same receptor but called it TRPM8.

Together with the two heat receptors, CMR1 can provide a physiological thermometer for most, but not all of the temperature range from 8 °C to around 60 °C. It is not yet clear whether there are further temperature receptors to be discovered or whether some as yet unknown mechanisms might modulate the range of the three we know. Similarly, the details of how the signals from these receptors are further transduced and processed remain to be explored. The finding that they are all from one family and respond quite similarly may contribute to the understanding of cross-reaction phenomena, e.g. the fact that noxious cold may feel like a burn, and the paradoxical burning pain evoked by simultaneous contact with warm and cool surfaces. With the discovery of the key molecules, researchers have finally found a handle to grab the burning issues of temperature sensation by.

(2001)

Further Reading

D.D. McKemy *et al.*, *Nature*, 2002, **416**, 52.
A.M. Peier *et al.*, *Cell*, 2002, **108**, 705.

What Happened Next

Subsequent research revealed further candidate receptors, including one for extreme (noxious) cold. In order to clarify the role of their receptor, which is now universally called TRPM8, Julius' team created a strain of knockout mice in which the gene for TRPM8 is mutilated to an extent that the membrane protein is produced but cannot function as a receptor. Simultaneously, two other groups independently used the same approach.

All three teams, publishing simultaneously in 2007, reported that the lack of TRPM8 suppresses the rodent's reaction to moderate cold. In cell cultures of knockout mouse neurons, the cold reaction is also suppressed, while the heat response functions normally.

The knockout mice turned out to be perfectly normal, except for their response to temperature differences. Normal mice prefer a surface of 30 °C to a colder surface, even if the colder one is at 25 or 20 °C, as researchers can show with very simple experiments measuring the time the animals choose to spend exploring two otherwise identical surfaces. Mice without a functioning TRPM8 molecule showed no preference between plates at 30 and at 20 °C. Only after the researchers cooled the colder area to 15, and then to 10 °C, did the animals develop a slight and a strong preference, respectively, for the warmer area. Only at 5 °C did their behavior fall back in step with that of the normal mice.

Thus, the knockout studies show, in agreement with molecular and cell biological investigations, that the bio-thermometer of the mouse – presumed to be very similar to our own – distinguishes between two types of cold. There is a slightly unpleasant chill, for which TRPM8 is the main sensor, and then there is dangerous cold, with temperatures close to the freezing point, for which there are other receptors. Details of the mechanisms involved and of the overlap phenomena between these regions of the temperature scale remain to be explored.

Further Reading

D.M. Bautista *et al.*, *Nature*, 2007, **448**, 204.
A. Dhaka *et al.*, *Neuron*, 2007, **54**, 371.
R.W. Colburn *et al.*, *Neuron*, 2007, **54**, 379.

Replicators Lose their Inhibitions

One of the coolest nanotech inventions ever was the development of molecular nanotubes from peptide rings designed to stack up all by themselves. As I've described that one in some detail in *Travels to the Nanoworld*, I'll include a different peptide technology here, which comes from the same laboratory, namely stick-like peptides that catalyze their own assembly from two shorter pieces.

If you want to get two molecules to react, positioning them in the right way could get you there quite efficiently. Using this simple chemical insight and some clever strategic thinking, Reza Ghadiri and his group at the Scripps Research Institute (La Jolla, California) created a completely new research field. Starting from an existing biological structure, the coiled coil (two alpha helices wrapped around each other), they declared one strand of it the template which helps the positioning of the other one, which we could call the target strand. They cut the target strand in halves and activated the ends so that the halves can react to form the whole. Then they demonstrated that the template can serve to position the halves, thus speeding up their reaction to form the complete target strand.

In this way they created a novel artificial enzyme, namely a peptide ligase, which can form a long peptide from two short ones. Moreover, if the target is identical to the template, the reaction is producing more templates and is therefore autocatalytic. In other words, the researchers created the first self-replicating molecule.

Since then, Ghadiri's group has developed this molecular toy further, to demonstrate how a simple chemical system can develop some elementary properties of life, such as replication, selectivity for chi-

The Birds, the Bees and the Platypuses. Michael Gross
Copyright © 2008 WILEY-VCH Verlag GmbH & Co. KGaA, Weinheim
ISBN 978-3-527-32287-9

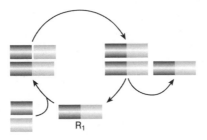

Figure 21 Self-replicating peptides.
(New Scientist, 24. 4. 1999)

rality, error correction, and interactions between overlapping reaction cycles in a complex hypercycle.

The group of Jean Chmielewski at Purdue University developed a similar system of their own, and incorporated switch functions sensitive to pH or other chemical parameters. Their peptide E1E2, for instance, contains an acidic "stripe" of glutamic acid residues along one side of the helix. At neutral pH, these side chains are highly charged, which makes the peptide chain prefer an unfolded conformation. At low pH, in contrast, the peptides are mostly helical and can serve as templates to guide their own synthesis from the fragments E1 and E2.

The trouble with this approach, however, is that once the ligation has taken place, the product binds even more tightly to the "enzyme" than the original fragments did. The system suffers from a problem known as product inhibition. This is why its output does not increase exponentially with time, as one would expect with a truly autocatalytic system.

To overcome this fundamental problem, Chmielewski's group reduced the stability of the coiled-coil complex by as much as they could without endangering the binding of the fragments necessary for positioning. They achieved this goal by shortening their self-replicating peptide E1E2 to a length of 26 residues, which they believe to be the minimum for the desired reaction to occur. Studying the self-replicating capacity of the new peptide, called RI-26, they observed a catalytic efficiency (catalyzed rate constant/unanalyzed rate constant) of 100,000, which is more than 20-fold higher than the previous record for self-replicating molecules. This efficiency approaches the range found in natural enzymes.

From these results, and other kinetic parameters observed, the researchers conclude that they have for the first time succeeded in breaking the deadlock of product inhibition in self-replicating peptides. While the group led by Günter von Kiedrowski at the University of Freiburg, Germany, has created self-replicating nucleic acids and bypassed this problem by temperature cycling (essentially as in the polymerase chain reaction), the destabilization of the template–product complex is clearly the most promising way towards a truly self-replicating system that can operate consistently in constant environmental conditions.

The availability of such peptides provides a unique opportunity to study complex molecular "behavior" in a simple system. Some of the processes involved will be spookily reminiscent of things happening in the living cell. For example, Ghadiri's group reported the emergence of "symbiosis" in their early work on peptide hypercycles. However, one should not be tempted to transfer these findings to the still largely mysterious field of the origin and pre-cellular evolution of life. It has to be stressed that biological molecules do not replicate themselves, but rather replicate each other. Furthermore, most researchers see RNA as a more promising candidate for the principal role in the early molecular stages of evolution. Thus, self-replicating peptides may have little to teach us about the roots of the tree of life, but they do add some interesting new branches to the tree of chemistry.

(2002)

Further Reading

D.H. Lee et al., Nature, 1997, **390**, 591.
A. Luther et al., Nature, 1998, **396**, 245.
R. Issac et al., Curr. Opin. Struct. Biol., 2001, **11**, 458.
R. Issac and J. Chmielewski, J. Am. Chem. Soc., 2002, **124**, 6808.

What Happened Next

This field has gone spookily quiet in the last few years, but I am hoping it will return to the limelight soon.

Biotronics: A Collision of Continents

In 2002, I co-authored a book about how the worlds of electronics and biology begin to merge, as researchers learn how to connect transistors to nerve cells, or artificial retinas to the visual cortex. The book was in German and soon ran into trouble, as a company claimed trademark rights over the German equivalent of the word "biotronics." So for a couple of years, until this legal tangle was sorted out, the following piece, written for Bio-IT World as a spin-off from the book, remained the only exponent of the biotronics idea in the public domain.

Biology is a rich and diverse science, branching out in many directions and sharing interdisciplinary research areas with many other disciplines. Along with the old and venerable enterprises of biochemistry and biophysics, there are new or newly fashionable additions, such as astrobiology and sociobiology. An increasing number of new arrivals seem to settle at the interfaces that biology shares with electronics and information technology. I should like to argue that this development is more than just a fashion trend, and that it reflects an actual merger process between the worlds of biological and technological information processing. We are observing a collision of continents that will make separations disappear and create entirely new landscapes.

In the traditional world view we have inherited over many generations, technological and natural systems are exact opposites, and mutually exclusive. People have always built tools that took some inspiration from nature, from levers through to cameras, but technological devices were bound to achieve their goals in different ways from their natural counterparts. To this day, no computer works like a brain, no

 The Birds, the Bees and the Platypuses. Michael Gross
Copyright © 2008 WILEY-VCH Verlag GmbH & Co. KGaA, Weinheim
ISBN 978-3-527-32287-9

microphone like an ear, and no camera like an eye. In our cultural tradition, people who violated this boundary – the most famous being Dr Frankenstein – had to be punished.

Current developments, however, are gradually nibbling away at this boundary, and it looks as though in ten years it will have become entirely meaningless. The separations are eroded from both sides. Technology is taking more inspirations from cellular biology than ever before. The entire field of nanotechnology is based on the insight that the cell's machinery works on the nanometer scale. Essentially, nanotechnologists ask: why shouldn't we be able to build machines like that too? In driving the length scale of technology down towards the cellular scale, we can follow nature's lead more or less closely. For instance, we can borrow ready made machines, such as molecular motors, from the cell and use them for our purpose. Alternatively, we could just use the principle of building machines out of proteins, but design our own.

And even if we bring our own (artificial) molecules, such as dendrimers or rotaxanes, we are still using biological construction principles, such as the ideas of self-assembly, weak (non-covalent) interactions, and modular design, creating complexity step-wise from simple building blocks. In any case, future technologies that are going to realize the potential of the nanoscale are bound to contain some elements of biological origin.

On the other side of the traditional divide, the understanding of what goes on inside the living cell has grown explosively over the last half-century. Some of the aspects that have emerged on the nanoscale are similar to our old large-scale technology. The cell contains linear and rotary motors, molecular assembly and disassembly lines, pumps, switchboards, and many other useful things. Some even have solar cells, light bulbs, or clocks. In the ways it stores and processes information, every cell has some features of a very small computer.

The more we understand the machine aspects of the cell and learn to develop the use of biological elements in our technology, the easier it becomes to create new interfaces between the two. At present, it is already possible to implant a primitive artificial retina into the human eye and connect it to the nerve system of the recipient in a way which actually allows them to see in a crude way. In a recent, rather controversial piece of research, scientists at the State University of New York managed to plug a remote control into the central nervous system of a rat, creating a "roborat" whose movements they could guide by direct communications from computer to brain.

While the fear of Frankensteinian misuse is never far away when wires are attached to living cells, there are many undoubted benefits that could arise from the ability to create interfaces between nerve cells and conventional electronics. Paraplegics could learn to walk, deaf people hear, and blind people see, if only they could be wired up appropriately.

Outside the body, the trend of "wearable computers" has produced IT equipment that can fit into spectacles or clothing. While this field has so far remained a playground for geeks, it is not difficult to foresee that some really useful medical applications could result as soon as affordable wearables (no matter whether they are worn under or over the skin) are made to interact with human physiology in a meaningful way. One promising candidate would be a combined glucose sensor/insulin dispenser for people with diabetes, which could be available in a matter of years.

In information technology, benefits from the merger with biology are already with us, and more of them are expected in the near future. The current explosion in genomic information would not have been possible without today's computers, and it even drives the development of IT in some areas. On the other hand, the exponential growth of computer performance predicted by Moore's law is bound to hit the final road block some time within the next decade, and biomolecules are among the promising candidates for further improvements.

In the not too distant future, patients may be fitted with barely visible medical appliances, which may use biomolecules as sensors, then some traditional electronics for information processing, and finally chemistry for actuation. The whole will be optimized such that the recipient's immune system will not recognize it as a foreign body. At that point, it will no longer be meaningful to define which parts of such devices will be biological, and which technological. Two continents will have become one.

(2002)

What Happened Next

In 2006, researchers managed to connect nanowires with nerve cells, as I will describe below (page 229).

The One-atom Quantum Computer

Back in the days at the Oxford lab, I used to share a computer room with Jonathan Jones, a colleague who, apart from his involvement in the staple business of the lab, also ran a sideline that appeared to involve calculations like 1+1. One day, I asked him what that was about and he explained to me that he was developing a quantum computer based on small organic molecules studied in an NMR machine (of which there were lots around). Thanks to Jonathan's tuition, I understood a minuscule part of the whole quantum computer business, enough to write occasional articles about it. They were all great fun, but I chose the following piece because it appeared with my favorite ever standfirst:

One calcium atom minus one electron makes one quantum computer. Michael Gross checks the sums.

Theoretical physicists have written many articles and entire books about the wonderful things one could achieve with a quantum computer. Many important computational problems are currently classified as "intractable" because the computation time rises at least as an exponential function of the number of variables, so the time becomes astronomical very quickly. A large quantum computer will be able to solve such problems, thanks to the special properties which allow one quantum-mechanical bit (qubit) to populate the 0 and 1 state at the same time. There is the slight disadvantage that all the encrypting methods used today in many fields from online banking to military communications will no longer be secure, but in exchange for that, quantum cryptography will give us a new kind of encryption

The Birds, the Bees and the Platypuses. Michael Gross
Copyright © 2008 WILEY-VCH Verlag GmbH & Co. KGaA, Weinheim
ISBN 978-3-527-32287-9

whose security is guaranteed by the laws of physics. So everything will be just fine. That's the theory, at least.

Over here, in the real world, quantum computers have now (2003) reached a crucial step in their very early development, comparable in computer history to the point at which the transistor was invented. Real world qubits have been produced in several different ways, and now the challenge is to get them to do something that deserves the name of calculation. Five years earlier, pioneering work using commercial NMR spectrometers to observe and manipulate the nuclear spins of certain atoms in small molecules succeeded in transferring quantum algorithms from theory to experiment.

It soon became obvious, however, that this technique is not suitable for being scaled up to the level of thousands of qubits, where a quantum computer could become useful and demonstrate its advantage over conventional computers. This is why experts gathered at a Royal Society discussion meeting in November 2002 were quite excited to hear from Rainer Blatt that his group at the University of Innsbruck, Austria, had managed to implement a quantum algorithm with two qubits, using a single trapped calcium ion as the hardware. While the NMR methods rely on observations of molecular ensembles floating around in an ordinary NMR tube, the Innsbruck work represents the first quantum calculation implemented on a system that is completely under the experimenter's control.

It is far from trivial to think of a computational problem that can be solved using only two bits of information. Even trickier, to think of a problem of this size that can be solved with fewer steps on a quantum computer than on an ordinary computer. There are only a few such problems around, and only a handful of simple quantum algorithms that researchers try to implement in their quantum computers. The Deutsch problem is probably the best-studied of these, and it has been used in the pioneering works of both NMR and ion-trap quantum computing. The task is, essentially, to determine whether two things are the same or different, assuming each can only be in one of two states.

Imagine you want to check whether the lights in your bedroom and in the bathroom are in the same state (e.g. both on). You would have to go to both rooms, notice the state of each light, and then compare the results. Now you could argue that this is too much trouble, because you're only after one bit of information. As the lights can be in the same state or in a different state, there are only two possibilities, so the answer to the question has the information content of one bit. But in classical physics, there is no way around the fact that you have to collect two bits of information (the state of each of the lights, which you aren't really interested in), compare the two, and derive your one-bit result from this comparison.

The Oxford physicist David Deutsch – a pioneer in quantum computation and popularizer of the multiple universes interpretation of quantum mechanics – showed, in contrast, that a quantum computer could determine such a result in one step without collecting the unwanted bits of information. One quantum algorithm that can perform such a "same or different" calculation is the Deutsch–Josza algorithm, which Blatt's group has now implemented using a single calcium ion. The Innsbruck researchers trapped a $^{40}Ca^+$ ion using laser cooling (in a device known as a Paul trap), and then used its vibrational motion and the excitation state of the single electron left over in the outer shell (which would be removed in the more familiar Ca^{2+} ion, the one that is good for your bones and teeth) as the two qubits. To address these, they used laser pulses of well-defined wavelengths. To improve the control over their system, they also used methods involving composite pulses, which were first developed for the radiofrequency pulses used in NMR.

The reliability and relative robustness that Blatt's group observe with this single-ion system suggest that in the near future it can be scaled up to comprise several atoms, possibly also of different species. The way from controlling one atom to controlling many may be long, but if and when quantum computers with large numbers of qubits become accessible, the implications will be truly revolutionary, and not just in theory.

(2003)

Further Reading

C.H. Bennett and D.P. DiVincenzo, *Nature*, 2000, **404**, 247.
S. Gulde *et al.*, *Nature*, 2003, **421**, 48.

What Happened Next

Progress has generally been slow in the quantum computation field, but in September 2007, real-world quantum computers came a step closer. Two research teams reported that single photons can be produced in a microchip and exchange quantum information through superconducting circuits. These findings combine to what the cover of *Nature* (27.9.2007) refers to as a "quantum bus." As it is based on established nano-fabrication methods, this achievement should be relatively straightforward to scale up to multiple qubits, and possibly even to mass-produced devices.

Further Reading

M.A. Sillanpää *et al.*, *Nature*, 2007, **449**, 438.
J. Majer *et al.*, *Nature*, 2007, **449**, 443.

Twist and Twirl

> As somebody who has worked with biomolecules for many years, I am all too familiar with the frustration resulting from the fact that these molecules are usually invisible and untouchable. Everybody involved with proteins or nucleic acids must have wished at some point to be able just to grab the two ends of a biopolymer and to pull it apart, or to twist it this way or that. A few years ago, this dream turned into reality.

Take a length of rope in both hands and twist it. If you turn in the same direction in which the individual strings of the rope are wound around each other, there comes a point where the rope cannot be wound any tighter and it will try to curl up into shapes that are twisted on a larger scale. If you turn the other way, the fibers of the rope will detach from each other at least in one place, opening up a loop.

Now imagine you had extremely small tweezers that enabled you to do the same kind of twisting and twirling with the rope that is double-stranded DNA. As Vincent Croquette reported at a 2003 workshop meeting of the EPSRC-funded Nanonet, his group at the CNRS laboratory for statistical physics in Paris has in fact optimized the methodology known as "magnetic tweezers" to allow them to do just that. Essentially, the researchers glue one end of a DNA double helix to a solid support, and the other end to a magnetic bead which they can move and turn around in an inhomogeneous magnetic field.

Making a bead spin to twist and untwist a molecular rope looks like just another demonstration of the amazing things one can do on the nanometer scale these days. But this one has the extra benefit of being useful in a number of ways, as the interest from molecular biologists shows, who are keen to apply the magnetic tweezers to their

The Birds, the Bees and the Platypuses. Michael Gross
Copyright © 2008 WILEY-VCH Verlag GmbH & Co. KGaA, Weinheim
ISBN 978-3-527-32287-9

specific problems in DNA replication and transcription. Essentially, all natural processes which involve DNA editing or readout require some extent of twisting. Being able to monitor both the rotation and the forces involved, researchers can gain unprecedented insights into these reactions. "The most exciting result" says Croquette, "is that you see an enzyme working in real time."

Croquette has already collaborated with the group of his CNRS colleague G. Charvin on DNA topoisomerases (enzymes that untwirl DNA by introducing a temporary cut in one strand through which the other strand can pass) and with the group led by Giuseppe Lia at the University of Milan on the Gal repressor, which blocks transcription by tying DNA into a loop. The single-molecule studies showed that a slight untwirling of the double helix favors the binding of a histone-like protein known as HU, which then bends the weakened part of the helix such that two molecules of the repressor can close the loop.

At the workshop, Croquette presented new work on the helicases Uvr-D and Rec-Q. These are enzymes which can bind to a loose end sticking out of a double helix and unravel the structure from there. Typically, they stay bound to one defined DNA strand, moving along from the 3' to the 5' end, i.e. opposite to the normal reading direction. Although such enzymes are essential for several different kinds of DNA repair and for the replication of a number of plasmids, very little was known about their function in molecular detail. In a subsequent piece of research, Croquette and his colleagues have described in detail the mechanical action of the helicase UvrD.

At the same workshop, Ralf Seidel of the Technical University Delft (Netherlands) presented another project using magnetic tweezers, which is a collaborative effort between the laboratories of Cees Dekker at Delft (famous for his work on carbon nanotubes) and Keith Firman at the University of Portsmouth, England, who organizes the Nanonet workshops. These researchers are studying type I restriction endonucleases, a kind of DNA editing enzymes which are remarkable for the long distance between the DNA sequences they recognize and those that they actually cleave.

Between these two events, the enzyme must travel a length of up to several thousand base pairs along the double helix, following its helical twist. Conversely, this implies that an immobilized enzyme of this type can move a DNA double strand by the same length, while also rotating it. Using the magnetic bead approach, the researchers can now both record this movement and apply a counterforce to measure how strong and efficient the endonuclease is as a molecular motor. They measured a translocation speed of over 500 base pairs per second.

But beyond the benefit of measuring physical data on a molecular scale, there is also the prospect of practical applications. Cees Dekker describes it as "the first step toward a biologically based actuator. Such a device," he says, "could link the biological and silicon worlds."

For anybody considering using the technique on their own molecules, the future is bright. "We are presently collaborating to develop a commercial magnetic tweezer apparatus," says Croquette. While the measuring technique usually requires the help of a physicist in the beginning, he promises that it soon becomes routine, and "the biological skill is the most critical aspect." Which means, if you can immobilize your molecule and stick a bead to the other end, you should also be able to twist it.

(2004)

Further Reading

G. Charvin *et al.*, *Proc. Natl Acad. Sci. USA*, 2003, **100**, 9820.
G. Lia *et al.*, *Proc. Natl Acad. Sci. USA*, 2003, **100**, 11373.
Nanonet UK: www.nanonet.org.uk

What Happened Next

I haven't heard any recent news from this field, but watch this space.

Multi-purpose DNA

"Misuse of DNA" is one of the areas which I covered repeatedly from the very beginning, which I think was Ned Seeman's DNA cube. Initially, this looked like sheer playfulness (see page 43 for impressions of the "crazy" phase of this field), but very soon some things with very real potential emerged from this playing, as these examples demonstrate.

DNA Computers Go Medical

In 1994, computer scientist Leonard Adleman surprised the world by demonstrating the first DNA-based computation. While his first calculation was only a proof of principle, a simple task performed with an inordinate amount of benchwork, it soon became clear that DNA computers might one day become really useful if they could be fully automated and reduced to the scale of the cell. Ten years later, the group led by Ehud Shapiro at the Weizmann Institute has created a DNA automaton that can "diagnose" symptoms of cancer and administer a "therapy" – at least *in vitro*.

In some cancers, including most prominently prostate cancer, routine diagnosis is already based on molecular signatures rather than anatomical anomalies. Using the gene expression levels that doctors commonly use to recognize this cancer, Shapiro's group designed a computational DNA molecule that can deal with a series of five yes/no questions in order to establish whether the typical markers of prostate cancer are present or not. Essentially, the molecular computer needs five positive replies (i.e. five markers to be present) to be able to release the drug.

The Birds, the Bees and the Platypuses. Michael Gross
Copyright © 2008 WILEY-VCH Verlag GmbH & Co. KGaA, Weinheim
ISBN 978-3-527-32287-9

The molecular computer is a long DNA hairpin (i.e. a double helix consisting of a single molecule bent back onto itself) containing the drug molecule (a short strand of DNA that interferes with gene regulation) in its bend. The double-stranded stem of the hairpin contains five "locks," each of which can be opened and removed by specific "keys," namely by the presence of the diagnostic messenger RNA in the amounts surpassing a defined threshold. When all five locks have been opened, the drug is released.

The researchers successfully applied this automaton to a test system that recreates the typical molecular signatures of prostate cancer *in vitro*, and then went on to carry out a similar "treatment" on a test tube model of small cell lung cancer. Nevertheless, Shapiro remained cautious in his press interviews. "It may take decades before such a system operating inside the human body becomes reality," he said. Apart from the concern of how well the molecular computer would survive in the body, the very process of introducing genetic material into a (possibly healthy) person would need to be considered carefully. Unlike the test tube with the cancer models, the body may contain edited mRNAs or protein factors that bind to the DNA computer in ways that cannot easily be predicted, even when the entire genome sequence is cross-checked beforehand.

Nevertheless, this work took the non-natural applications of DNA, which have spent ten years in a molecular playground, back to the real world and to the prospect of real usefulness. In a similar development from the field of artificial 3D structures built from DNA, Gerald Joyce introduced clonable DNA strands that fold up to form octahedra all by themselves. Bringing together the structural and computational powers of DNA with its natural efficiency to work and replicate on the single molecule level bears immense promise for nanotechnology and medicine.

Figure 22 The DNA doctor. Shown here is a simplified version with just two locks that need to be opened before the DNA drug (the bend of the hairpin) can be released. A specific key for each lock binds to the short sequence of single-strand DNA at the end. By completing the double helix, the key also completes a recognition site for a specific restriction enzyme that will cut the double helix in a well-defined way, opening access to the next lock.

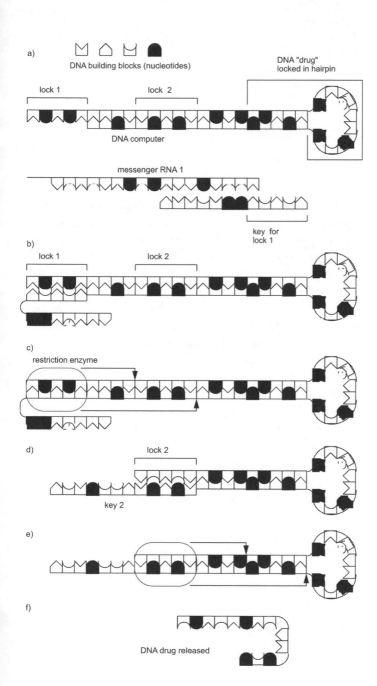

a)

DNA building blocks (nucleotides)

DNA "drug" locked in hairpin

lock 1   lock 2

DNA computer

messenger RNA 1

key for lock 1

b)

lock 1   lock 2

c)

restriction enzyme

d)

lock 2

key 2

e)

f)

DNA drug released

Further Reading

Y. Benenson *et al.*, *Nature*, 2004, **429**, 423.
W.M. Shih *et al.*, *Nature*, 2004, **427**, 618.

DNA Gets Hands And Feet

DNA used to be the information specialist that left the action to the proteins, or possibly some RNAs. Not any more. Thanks to some cleverly designed DNA sequences, the genetic material can now form complex architectures, electrical wires, molecular computers, and now even machines with moving parts. Following on the heels of the development of an automatic medical device based on DNA (above), two papers showed that DNA can have hands to hold and release things and feet for molecular walking tours.

The group led by Friedrich Simmel at the University of Munich, Germany, designed a DNA "hand" that can be instructed to cyclically grasp and release a molecule of the enzyme thrombin. The device is based on a 15-base DNA sequence, or aptamer, known to bind thrombin. "We chose the thrombin aptamer because it was short, well-characterized and had a low dissociation constant," says Simmel. He combined this element with a 12-base tag that makes the hand controllable by additional DNA sequences. Adding a specific DNA molecule (Q) that recognizes this tag, the researchers could dislodge the bound enzyme. Another DNA sequence (R) outcompetes the "hand" in its interaction with Q, and thus sets it free to bind the enzyme again.

Meanwhile, William Sherman and Ned Seeman (famous for constructing 3D objects from DNA) at New York University have built a walking DNA robot using a similar approach. Its two feet have different sequences, which specifically recognize two kinds of foothold on the DNA track it walks on. Soluble DNA strands added to outcompete the binding interactions allow the researchers to get the DNA robot walking in a controlled manner.

Simmel recognizes that his "hand" and Seeman's "feet" converge naturally: "In combination with DNA walkers," he explains, "our device could form the carrying part of a DNA motor which grabs a molecule in one place and releases it in another. Similar devices could be constructed which bind or release non-biological molecules rather than proteins." Further research from Simmel's laboratory even shows that DNA machines can be controlled biologically, via the cell's transcription process. Clearly, with DNA robots evolving at that pace, it cannot be long before they start to talk and serve drinks.

(2004)

Further Reading

W.U. Dittmer *et al.*, *Angew. Chem. Int. Ed.* 2004, **43**, 3550.
W.B. Sherman and N. Seeman, *Nano Lett.*, 2004, **4**, 1203.
W.U. Dittmer and F.C. Simmel, *Nano Lett.*, 2004, **4**, 689.

What Happened Next

In 2006, researchers from the DNA origami project at CalTech reported two-dimensional images created simply by designing DNA sequences. Smileys made of DNA made the cover of *Nature* (16.3.2006).

Marveling at Diatoms

During my years as a postgraduate and postdoctoral researcher
I have seen many many undergraduates floating past. Most of
them have never shown up on my radar again, but here is one
who has continued to surprise me by publishing really cool research papers. When I was in the first year of my PhD at the University of Regensburg, Germany, Nils Kröger joined the Jaenicke
group for a short practical course, and then shocked everyone by
choosing a different lab for his diploma and PhD. He flourished
spectacularly, though, practically making the field of biochemical
analysis of diatom shells his own. While I would have covered
his papers even if they had come from someone completely
anonymous, it was kind of heartwarming to write about these
stories with the knowledge that this brilliant work came from
someone I knew before he even had a degree to his name.

Build Your Own Diatom Shells

Diatoms are single-cell algae represented by thousands of species
in oceans and freshwater lakes around the globe. They form shells
with beautiful patterns of sub-microscopic pores, which have posed
major challenges to biomineralization research. As the patterns are
genetically determined, there are evidently biomolecules involved, but
these are so different from normal proteins that it took researchers
many years even to extract them out of the shell material.

Nils Kröger and his coworkers at the University of Regensburg,
Germany, have previously isolated several different peptides (called
silaffins) and long-chain polyamines from diatom shells and shown

 The Birds, the Bees and the Platypuses. Michael Gross
Copyright © 2008 WILEY-VCH Verlag GmbH & Co. KGaA, Weinheim
ISBN 978-3-527-32287-9

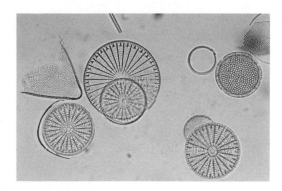

Figure 23 Electron micrograph of diatom shells. The intricate, lace-like patterns of nanometer-sized pores are genetically de- termined. Just how these patterns emerge is one of the trickiest questions of modern biology.

that they catalyze silica precipitation. In 2002, the group reported that post-synthesis modifications, including polyamine extensions attached to lysine residues and phosphorylated serines, are crucial for the function of silaffins *in vivo*. When the researchers first extracted silaffins from diatoms using hydrogen fluoride, these modifications had been lost. Now they have restored the full, "native" versions of all three silaffins found in *Cylindrotheca fusiformis* and established a recipe that predicts how these molecules must be combined in order to obtain diatom-like nanostructures in the test tube.

Kröger and co-workers found that native silaffin-2 on its own, a 40-kDa phosphoprotein, has little effect on the formation of silica precipitates. But together with native silaffin-1A (6.5 kDa) it appears to have a regulatory function, as it can stimulate or inhibit the other protein's activity, depending on the relative concentrations. The minimal construction kit required to create silica structures with the right pore size (100–1000 nm) *in vitro* must contain one cationic component, such as a long-chain polyamine or the native silaffin-1A with its polyamine modifications, as well as one anionic component such as the highly phosphorylated native silaffin-2. The authors propose that these two crucial components form a self-assembling matrix based on electrostatic interactions. In all the combinations studied, the cationic molecule catalyzed the silica formation, while the anionic one regulated the process.

If this recipe is more widely applicable, it will allow researchers to try out an infinite number of molecular variations on the theme, thus opening the way to nanostructured synthetic materials based on the diatom's construction principles.

(2003)

Further Reading

N. Kröger et al., Science, 2002, 298, 584.
N. Poulsen et al., Proc. Natl Acad. Sci. USA, 2003, 100, 12075.

Nanotechnologists Employ Algal Architects

Diatoms are unrivalled architects at the nanoscale. The silicon shells of these single-cell algae are porous, highly symmetrical boxes with species-specific, lace-like patterns in nanometer dimensions. Similar control of nano-structuring processes would be useful in many areas of nanotechnology, but mimicking the diatom's construction work has proven a major challenge. Now, researchers in the US have found that the natural diatom shell can be used as a foundation for artificial nano-constructs, a much more accessible route than biomimetics.

Chad Mirkin and colleagues at the Northwestern University in Evanston, Illinois, stripped diatoms down to the pure inorganic skeleton by immersing them in an acid bath aptly named the "piranha." This treatment left the shells susceptible to functionalization with an amino-silane reagent. The free amino functions were in turn used to couple DNA oligonucleotides to the diatom shells. Using complementary DNA strands coupled with gold particles (13 nm) in order to make the coating easily visible in the electron microscope, the researchers demonstrated that the entire surface of the empty shell can be functionalized and covered in artificial materials. Moreover, the researchers could add up to six further gold layers onto the first one.

The procedure worked equally well with two different diatom species, *Synedra* and *Navicula*. Thus, researchers might one day choose any of the thousands of different diatom species for their specific purposes. For example, to create a particular nanoscale pattern in gold (or any other) particles, one would only have to find a matching diatom shell and build on that. Alternatively, one could "easily modify the diatom surfaces with many different functional groups," suggests Mirkin. For example, one might produce microscopically small DNA arrays in large copy numbers.

(2004)

Further Reading

N.L. Rosi *et al.*, *Angew. Chem. Int. Ed.*, 2004, **43**, 5500.

What Happened Next

Understanding and mimicking diatom shell architecture has remained one of the biggest and most interesting challenges at the biology/nanotechnology interface. In March 2007, Kröger's team reported immobilization of a protein on the silica shell of a diatom. Later that year, the group of Börner at the Max-Planck Institute of Colloids and Interfaces at Golm, Germany, described a bio-glass fiber created from diatom silica and peptides.

Nature's Warning Signs

Sometimes, an experiment so simple that any primary-school child could have designed it and, indeed, even carried it out provides an important piece of information about how our world works. One of these really simple but brilliant experiments is reported in the following cool story:

In the laboratory, all bottles containing toxic substances must carry a label with a warning sign that is recognized globally. In nature, different species that produce toxins to make themselves unpalatable to predators are often found to share universal warning signs, such as conspicuous, brightly colored patterns. While the uniformity in the laboratory is due to legal regulations and serves the safety of the person who might ingest the poison, the natural phenomenon known as Müllerian mimicry arises from the evolutionary advantage it confers not to the eater (predator) warned, but to the meal (prey) spared. If several species share the cost of the errors that predators make while learning to read the warning signs, each suffers less significant losses. With a very simple and elegant experiment, British researchers have now found an additional benefit to the sharing species.

John Skelhorn and Candy Rowe at the University of Newcastle fed chickens with red breadcrumbs treated with either quinine, or bitrex (the anti-nail biting stuff), or a combination of both, along with green breadcrumbs sprayed only with water. Within six trials, the chicks that were fed only one of the two bitter chemicals had learned to avoid the red crumbs. Surprisingly, the chicks exposed to a mixture of bitrex- and quinine-flavored crumbs learned their lesson in half as many rounds. Moreover, in a test taken four days after the training, those chicks that had been exposed to two different bitter substances,

The Birds, the Bees and the Platypuses. Michael Gross
Copyright © 2008 WILEY-VCH Verlag GmbH & Co. KGaA, Weinheim
ISBN 978-3-527-32287-9

were much better at avoiding the red crumbs than both single-flavor groups.

These results suggest that, on top of the "shared cost" benefit, the sharing of a warning sign between prey species with different toxins provides additional protection, as it makes the warning sign more memorable. Essential laboratory safety, as invented by Nature, a long time before all those black and yellow sticky labels!

(2005)

Further Reading

J. Skelhorn and C. Rowe, *Proc. Roy. Soc. Lond. B*, 2005, **272**, 339.

All on One Chip

Another simple but efficient experiment, though this time it's somewhat above school level.

Making just one version of two mirror-image molecules is one of the most important challenges that chemists face today. Chiral catalysis, i.e. speeding up the reaction path leading to the desirable version (enantiomer) but not the one producing its mirror-image, is a promising way to obtain the pure enantiomers that are needed in many applications, especially in the pharmaceutical industry. In 2006, chemists in Germany succeeded in combining chiral catalysis and analytical separation of the resulting enantiomers on a single chip. Their approach could be scaled up to serve in high throughput screening for new enantioselective catalysts.

The groups led by Detlev Belder and Manfred Reetz at the Max Planck Institute at Mülheim, Germany, created an integrated catalysis/analysis chip, which combines a reaction channel with a separation channel. The reagents are loaded into separate microvials, from where they are sucked into the long meandering reaction channel, which is designed to ensure their mixing and incubation, using either a vacuum or an electrical field. A miniaturized version of capillary electrophoresis known as microchip electrophoresis (MCE) serves to separate and analyze the products of the reaction. Belder and colleagues had previously shown that this technique can perform the separation of enantiomers at record-breaking speed.

As a prototype application for their new lab-on-a-chip system, the researchers analyzed the enantioselectivity of mutant epoxide hydrolase enzymes, which relate to Reetz's work on optimizing enzymes by artificial evolution. In the case of the wild type enzyme from the fun-

The Birds, the Bees and the Platypuses. Michael Gross
Copyright © 2008 WILEY-VCH Verlag GmbH & Co. KGaA, Weinheim
ISBN 978-3-527-32287-9

gus *Aspergillus niger*, the researchers could even use cell lysates and whole cells for the analysis. In addition to this application, the researchers say, "the device also has great potential for high-throughput screening."

Hans Niemantsverdriet from the Technical University Eindhoven, The Netherlands, agrees: "The work elegantly illustrates the potential of miniaturized catalytic devices for research purposes," he said. "I am convinced that small to medium sized catalytic devices also have a future in the production of specialty products, in particular if we learn how to integrate smart catalysts with measuring and control techniques on the microscale."

(2006)

Further Reading

D. Belder *et al.*, *Angew. Chem. Int. Ed.*, 2006, **45**, 2463.

What Happened Next

Further research from Belder's group, published in 2007, combined nanospray mass spectrometry with nanofluidic reaction and separation devices as described below. Thus, using a single chip placed in a mass spectrometer, the researchers are able to make molecules, purify them, and identify them by their molecular mass.

Platinum Stories

I don't know why, but platinum (Pt) is an element which I find intriguing. Maybe it is down to the fact that it is very noble, hence inert, but as a highly efficient catalyst it can make so many things happen. And I remember using a platinum electrode in the undergraduate practical course. The assistants weighed the thing before and after I borrowed it to milligram precision, so I guess the stuff must be really precious. For these reasons, I keep an eye on the journal *Platinum Metals Review* where I sometimes find the most amazing stories, including the following two:

Precious Platinum Photographs

An image of a moonrise over a lake, printed in platinum and then modified with the gum bichromate process, sold for nearly three million dollars at Sotheby's in 2006, setting a new price record for any art photograph.

The record-breaking print, entitled "The Pond – Moonlight," dates from 1904 and is the work of the American artist Edward Steichen (1879–1973). While there are three known prints originating from the same negative, the high degree of manual skill required to produce each print, along with the longevity guaranteed by the precious metal, helps to make such works highly desirable in the world of art.

"The platinotype process, invented in 1873 by William Willis of Bromley, Kent, was widespread around 1900, and highly esteemed as the finest method for making beautiful, permanent photographic prints," says Mike Ware, an expert in unusual photographic techniques. "Platinotype paper was available commercially and only slightly more expensive than the silver halide papers of the day."

226 *The Birds, the Bees and the Platypuses.* Michael Gross
Copyright © 2008 WILEY-VCH Verlag GmbH & Co. KGaA, Weinheim
ISBN 978-3-527-32287-9

In rare cases, such as the Steichen prints, the original black and white platinum prints were used as a substrate to which further layers with different tones and colors were added.

Steichen applied a solution of gum arabic, a pigment, and a dichromate to the print, then exposed the print through a negative once more. Light triggers the reduction of dichromate to Cr(III), which can cross-link the macromolecules of the gum, which then solidifies and traps the pigment. Using this process, several layers incorporating different pigments can be added, turning the original black and white print into a colorful – and valuable – work of art.

Further Reading

M. Ware, *Platinum Metals Rev.* 2005, **49**, 190.
M. Ware, *Platinum Metals Rev.* 2006, **50**, 78.

Platinum Rubles

During a period of nearly two decades in the eighteenth century, rubles made from (nearly) pure platinum were standard currency in Russia (and also the first coins ever to be made from this metal). They were withdrawn in 1846, when the availability of cheaper platinum from Colombia made the rubles an attractive target for counterfeiters. Experts at the companies Heraeus in Germany and Johnson Matthey in London have unleashed the whole arsenal of non-destructive analytical methods available today on a handful of surviving coins to establish how they were made and how the real ones can be distinguished from the fakes.

Heraeus owns a set of four coins (at face values of 3, 3, 6, and 12 rubles) as well as a medal commemorating the coronation of Tsar Nicholas I in 1826. Density measurements on these coins confirmed that the platinum used for these coins was less than pure. The main contaminants include the metals gold, iridium, rhodium, and iron. Depending on the ratio of iron over iridium, such coins are often ferromagnetic. The German researchers scrutinized the surface structures of the coins using both optical and scanning microscopies, and analyzed the interior using SQUID (superconducting quantum interference device) microscopy. Using these methods, the researchers established that the loose platinum "sponge" obtained from the dissolved phase must have been compacted by forging and rolling out before the coins were minted.

The company Johnson Matthey also holds four platinum ruble coins, whose provenance is subject to legends, but cannot be firmly established. Detailed investigation by magnetic permeameter and density measurements, as well as by scanning electron microscopy and X-ray diffraction revealed that two of these coins had a significantly higher purity than all authentic rubles known, and were thus in all likelihood forgeries (even though it takes an exceptionally stupid criminal to put more platinum into a forged coin than is contained in the real thing). The two remaining coins, however, an 1830 6-rouble coin and an 1835 3-rouble coin matched all the characteristic properties observed in other products of the Tsar's platinum mint and were thus deemed authentic.

(2006)

Further Reading

C.J. Raub, *Platinum Metals Rev.* 2004, **48**, 66.
D.F. Lupton, *Platinum Metals Rev.* 2004, **48**, 72.
D.B. Willey and A.S. Pratt, *Platinum Metals Rev.* 2004, **48**, 134.

Nanowires Plugged Into Nerve Cells

I know I keep banging on about biotronics and how the boundaries between electronic and biological systems are disappearing. But of all the work I've come across in this area, this particular example had the biggest wow factor, as in this case, for the first time, the electronic components were operating on the same length scale as the nerve cell they were plugged into. This really is the way forward.

The new age of biotronics, where electronic devices can directly and easily be interfaced with living organisms, has been inaugurated by the work of Peter Fromherz and others, who successfully connected electronic devices to individual nerve cells. The group led by Charles Lieber at Harvard, which has played a key role in the development of nanowire electronics over the last years, has now developed arrays of nanowire transistors which can contact not just individual neurons, but even individual dendrites and axons (outgrowths of the nerve cell) at multiple locations.

Lieber and his coworkers created arrays of silicon nanowire transistors and passivated the contacts to avoid corrosion during the extended incubation times needed for cell culture (up to ten days at 37 °C). Rather than rely on chance encounters between neurons and nanowires, they patterned polylysine films onto the substrate to define the areas where they wanted the neurons to spread. In typical experiments, they added the cells to the substrate, allowed one hour for attachment, washed off any unattached cells, and then incubated for four to eight days to allow the neurons to spread their axons and dendrites along the pre-defined paths leading towards the nanowire transistors.

Figure 24 The axon of a single nerve cell extending across an array of 50 nanowire transistors. Schematic draving based on microscopic images of the neuron and the array. The body of the neuron is at the bottom. Progress of a nerve signal along the axon can be followed in real time using the transistors. © PhotoDisc/Getly Images

Applying this approach in a range of different geometries, the Harvard researchers consistently obtained a high percentage (over 80%) of active contacts between the nanowires and the dendrites and axons. In one experiment, a single neuron was made to run its axon past an array of 50 transistors lined up at 10-µm intervals, 43 of which established functional contacts (Fig. 24). Such arrangements will enable researchers to study the electrical behavior of individual nerve cells in unprecedented detail and with minimal disturbance.

So far, these are just experiments in a Petri dish. However, considering the elegance, success rate, and size compatibility of this approach, it appears likely that within a few years, nanowires will serve to reconnect the neural pathways interrupted in blind or tetraplegic patients. The merger between biology and electronics is imminent.

(2006)

Further Reading

F. Patolsky *et al.*, *Science*, 2006, **313**, 1100.

Towards the Perfect Biosensor

Kevin Plaxco is an old friend from the early post-doc days at Oxford, and probably one of the two or three most brilliant scientists I know in my generation. In recent years (2003–6), he has developed some really cool sensors based on the ideas of protein folding. I have done a few stories on his sensors, of which this is the most comprehensive.

In the beginning, there was the folding problem. Many research groups around the world study the folding of proteins (or sometimes of RNA), sometimes with a biotech background or to get to grips with the biological systems, but often just for the intellectual challenge of working out how and why chain molecules fold up into a highly specific 3D structure with such remarkable speed and efficiency.

Kevin Plaxco had several years of folding research on his CV when he set up his own research group at the University of California at Santa Barbara. Looking for something more useful to do with his folding experience, he hit upon the idea of using folding processes for sensors. The first device his group developed was a gene sensor dubbed E-DNA, containing a folded-up probe sequence of DNA which will stretch out only when the exactly matching target sequence turns up. By stretching out, the DNA removes a ferrocene label fixed to its one end from the gold surface to which its other end is immobilized and thereby suppresses an electrochemical signal that was recorded previously.

DNA can be used for many different things (as explained above in the chapter about misuse of DNA), including molecular recognition of other kinds of targets. DNA molecules with highly specific binding properties selected from a large pool of random sequences are known as aptamers. Since the discovery of this concept in 1990, aptamers with many different specificities have been reported, which have been used in many applications including fluorescence-based sensors, known as aptamer beacons. In 2005, Plaxco's group, in collaboration with their campus neighbor and Nobel laureate, Alan Heeger, used one such aptamer, with specificity against the blood coagulation factor thrombin, to convert the E-DNA into an electronic aptamer-based (E-AB) sensor for this factor.

This first aptamer-based electrochemical sensor showed extremely high specificity, as the signal change could only be triggered by the DNA aptamer folding around the thrombin target, an event which is much more specific than simple adhesion processes. Like the E-DNA it had the advantage of being a reagent-less sensor, implying that this technology could be developed into a simple, hand-held device, which the user would only have to dip into the sample to be analyzed. "Making molecular beacons electronic has a huge impact on their convenience," Plaxco says. "And because the electrochemical background of a typical real-world sample like blood serum is many many orders of magnitude lower than its fluorescence background, electrochemical sensors can work in really really dirty samples."

However, both the thrombin sensor and the original E-DNA had the disadvantage of being "signal-off" sensors, i.e. the signal recorded in the absence of the target is weakened when target molecules are detected. This is where the competing group of Ciara O'Sullivan at the Universitat Rovira i Virgili at Tarragona, Spain, could go one better. They also developed an aptamer-based sensor for thrombin, but subtle chemical differences in the design of the aptamer sequence allowed their device to work in the more desirable "signal-on" mode.

Plaxco's group had to invest additional work in their E-AB sensor to enable it to work in the signal-on mode (Fig. 25), which led to an improvement of the sensitivity by a factor of ten. "By making our sensor signal-on," Plaxco explains, "we go from a 30% signal loss at saturation to a 300% signal increase."

The Santa Barbara group also developed a highly sensitive device to detect cocaine even in biological fluids and in environments where its presence has been masked deliberately. One of their most recent papers describes a sensor for PDGF (Platelet-Derived Growth Factor) which is sensitive enough to be useful in cancer diagnosis. Unlike the fluorescence-based sensors that other groups have developed for the

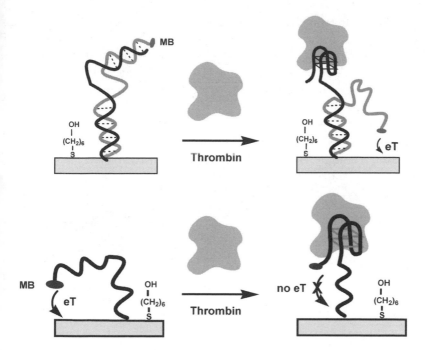

Figure 25 Examples of a signal-off (top) and a signal-on aptamer sensor against thrombin. In the signal-off sensor, binding of the thrombin molecule creates a secondary structure in the DNA aptamer, making it too rigid to approach the surface closely enough for electron transfer to occur. Conversely, in the signal-on sensor, the DNA starts out in a double-helical conformation. Recognition of the target molecule occupies one DNA strand, leaving the other one free to move and produce an electrochemical signal.

same purpose, this electrochemical device can be applied directly to blood serum, which only needs to be diluted two-fold to reduce the ionic strength of the sample.

(2006)

Further Reading

Y. Xiao *et al. Angew. Chem. Int. Ed.* 2005, **44**, 5456.
Y. Xiao *et al.*, *J. Am. Chem. Soc.* 2005, **127**, 17990
A.-E. Radi *et al.*, *J. Am. Chem. Soc.* 2006, **128**, 117.
B.R. Baker *et al.*, *J. Am. Chem. Soc.* 2006, **128**, 3138.
R.Y. Lai *et al.*, *Anal. Chem.* 2007, **79**, 229.

What Happened Next

In October 2007, the group led by Joseph Wang at Arizona State University, Tempe, reported an aptamer-based sensor system using two aptamers and reaching sensitivities in the femtomolar (10^{-15} mol per liter) range.

Further Reading

Y. Xiang *et al.*, *Angew. Chem. Int. Ed.* 2007, doi 10.1002/anie.200703242

Hairy Ball Theorem Untangles Chemical Problem

Nanoparticles are tiny specks of matter that measure only a few millionths of a millimeter across. They are typically well-rounded individuals – to an extent that it is difficult to involve them in any interactions that depend on geometrical preferences. Researchers at the Massachusetts Institute of Technology (MIT) at Cambridge, MA, reported in January 2007 that they had succeeded in breaking the symmetry of nanoparticles using a mathematical principle known as the hairy ball theorem. Which I considered a very cool way of putting the rather theoretical and often eccentric science of topology (the one that tells us that coffee mugs and doughnuts have essentially the same shape!) to practical use.

The hairy ball theorem states that on a sphere covered in fur one cannot brush all the hairs flat without creating at least two whirls, technically known as singularities. By contrast, a hairy doughnut-shape (torus) can be brushed flat without any problems.

German-speaking mathematicians call this the hedgehog theorem, as it also implies that a rolled-up hedgehog will be vulnerable in places. The theorem is useful to many people in different ways. For example, for meteorologists it predicts that as long as there are winds on our planet, there must be a cyclone or an anticyclone somewhere. Conversely, if you ever land on a planet that has winds but no cyclones, you will know immediately that it must be a giant doughnut. It goes without saying that the theorem is important for hairdressers, but how about chemists?

In their quest to use nanoparticles to build higher order structures the group led by Francesco Stellacci at MIT decided to brush up their

The Birds, the Bees and the Platypuses. Michael Gross
Copyright © 2008 WILEY-VCH Verlag GmbH & Co. KGaA, Weinheim
ISBN 978-3-527-32287-9

Figure 26 The hairy ball theorem predicts among other things that on a planet with winds, there must always be at least one cyclone. Unless the planet is a torus (donut shape), in which case the wind can go round and round without any cyclones.

math skills and apply the theorem to chemistry. Accordingly, they turned gold nanoparticles into tiny furballs by depositing self-assembling thiol monolayers onto their surfaces. Using transmission electron microscopy, the researchers could show that the thiols arranged themselves in parallel rings, like pins stuck into a globe along the latitude lines. The singularities manifest themselves as single molecules of thiol sticking out from each of the poles. Unlike the molecules in the rings, the polar ones are poorly stabilized by their neighbors, and thus they are easier to displace than the others.

The MIT group managed to replace the polar thiols with longer chain "handles" carrying carboxylic acid groups at the other end. Furthermore, they polymerized these constructs by applying the widely known reaction that is used in the production of nylon. They dissolved the coated nanoparticles in toluene and brought them into contact with an aqueous solution of 1,6-diaminohexane. They could then harvest the polymer of nanoparticles from the boundary between these two immiscible fluids, just like nylon.

Group leader Stellacci thinks that this research will become useful for both fundamental science and new applications. "We have created the nanoscale equivalent of polymers. We expect that these materials will show the validity of some of the fundamental assumptions of polymer physics and present a plethora of new properties," he explained.

Nanotech pioneer Cees Dekker from the Technical University of Delft, The Netherlands, is enthusiastic about the new development. "This is very original work that may open a new avenue for using nanoparticles," he said, adding that "it adds a new functionality to them, namely the ability to bond to other particles in specific directions, just like atoms make up molecules."

So the theorem that causes nothing but trouble for hedgehogs, hairdressers, and meteorologists, has actually proven tremendously useful for chemists.

(2006)

Further Reading

G.A. DeVries, *et al.*, *Science*, 2007, **315**, 358.

A Liquid Mirror for the Moon

I first learned about ionic liquids at a catalysis workshop in 2004 and was duly impressed. But I didn't imagine that scientists would one day discuss sending such liquids to the moon. Researchers have succeeded in coating an ionic liquid with a reflective surface, making it (almost) suitable for a moon-based telescope based on a parabolic mirror made of a rotating liquid. Operating down to a shivering 130 K, this telescope will be as cool as they come.

Cosmologists have long argued that a moon-based telescope with a parabolic mirror made of a rotating liquid would be ideally suited to study very distant (and thus very old) structures of the Universe in unprecedented detail. Chemists have now succeeded in creating a liquid-based system that comes very close to the requirements for such a project.

Ermanno Borra at the University of Laval, Canada, together with colleagues elsewhere in Canada, the US, and Northern Ireland, tested the properties of several different types of liquids and identified a commercially available ionic liquid as the most promising candidate.

Ionic liquids are essentially salts that remain liquid at ambient temperatures and sometimes even at very cold ones. Over the last few years, researchers have paired up many different types of charged molecules to form such liquid salts. They have shown promise in catalysis research. They are of particular interest as "green" solvents, as they don't evaporate (unlike typical organic solvents which tend to be volatile), which is also their key advantage for space-based applications.

The Birds, the Bees and the Platypuses. Michael Gross
Copyright © 2008 WILEY-VCH Verlag GmbH & Co. KGaA, Weinheim
ISBN 978-3-527-32287-9

Upon spray-coating the surface of the chosen ionic liquid, 1-ethyl-3-methylimidazolium ethylsulfate, with silver, the researchers found a much-improved reflectivity in the relevant infrared wavelength range compared to silver-coatings on other types of liquid, such as PEG (polyethylene glycol).

Analyzing the surface of the silver-coated ionic liquid, Borra and colleagues found colloidal particles in the size range of a few tens of nanometers. Hypothesizing that this was the reason why the reflectivity wasn't quite as good as in pure metallic silver, they changed their recipe to include a layer of chromium before applying the silver. The resulting double-coated liquid proved so good in reflectivity, the authors say, that the rest is a mere matter of technological fine-tuning.

Similarly, the melting point of the commercial ionic liquid used as a base isn't quite low enough to ensure the material stays liquid at the ambient temperatures of the moon, but the researchers express confidence that among the millions of possible combinations of ions, the optimal one for the lunar telescope can be identified.

But is this the future of astronomy or just pure lunacy? Paul Halpern, a physics professor at the University of Sciences, Philadelphia, and author of several cosmology books, welcomed the proposal enthusiastically. "It may indeed become economically feasible to establish a lunar telescope," Halpern said. He concluded: "For this, a giant rotating dish of highly reflective liquid could prove ideal. A reflector 100 meters in diameter could collect thousands of times more light than the Hubble Space Telescope and potentially image the primordial stars believed to have formed in the very early Universe."

(2007)

Further Reading

E.F. Borra *et al.*, *Nature*, 2007, **447**, 979.

Epilogue: The Next Fifteen Years

Looking through the articles I have written in the past 15 years to pick out the ones collected in this book, I realized how futile some of my optimistic predictions have proven. In the excitement over a new discovery, one gets easily carried away and extrapolates the latest development toward a rosy future. My standard prediction runs along the lines: now that the fundamental questions have been solved, practical applications should become possible within the next five (or ten?) years.

But have they? In some cases, my optimism turned out to be justified. Green Fluorescent Protein (GFP, pages 101 and 193) and RNAi (page 179) became widespread in laboratory procedures within a matter of months. In other cases, my outlook may have been too rosy, as unexpected roadblocks turned up at the next bend after the breakthrough discovery. Thus, gene therapy continues to be just out of reach, as it was in the mid-1990s. Bacteriorhodopsin and other biomolecules were considered exciting alternatives for computation, but the runaway progress of silicon chips, still obeying Moore's law to this day, has left these contenders behind. And the expression of spider silk proteins in goat milk (page 188) did not lead to the production of spider silk in useful quantities.

Prophecies have the nasty habit of falling back on the hapless prophet. Maybe my optimism isn't quite as embarrassing as those negative predictions that are often quoted with the benefit of hindsight, for instance that airplanes would never fly, and personal computers would never find a market. But still, one needs to be careful in predicting the unpredictable future.

Having said that, I can't resist the temptation to speculate a little about what stories I might include if I were to do a similar collection in 15 years' time. If, of course, books still exist in 2023. Their replace-

The Birds, the Bees and the Platypuses. Michael Gross
Copyright © 2008 WILEY-VCH Verlag GmbH & Co. KGaA, Weinheim
ISBN 978-3-527-32287-9

ment with e-books is one of the predictions I've read repeatedly over the years, without any signs of it actually coming true.

Current global problems, from overcrowding through to global warming, will not have gone away by then, but hopefully we will have come closer to constructive solutions. (There's my runaway optimism again – realistically, we should count ourselves lucky if we haven't wiped out the biosphere by then!) In a sense, the leading edge technologies, which I mostly write about, will not matter as much to the fate of humankind on a global scale as the robust technologies that can help the majority of people who still fight hunger and infectious diseases as their main enemies.

Vaccines that are affordable for the poorest nations, drugs that can be stored without the need for a fridge, or computers that can carry the information revolution to Africa, these are the things that will probably have a lot more impact over the next 15 years than any next-generation chip that allows your PC to crash even faster, or a drug that increases the life expectancy of rich people from 87 to 88. In other words, the fate of the world will depend critically not on how fast the leading edge of science and technology will move forward but on how well the trailing edge can catch up.

But in terms of science as an intellectually stimulating and fun activity, of course, much of the excitement is to be had at the leading edge. Bearing in mind that these advances will not make the world a better place, here are some crazy, sexy, and cool things that I would like to be able to include in my next round-up:

- A real-world quantum computer. Scrap the thought experiments. People will only start to believe that quantum mechanics is real when it's sitting on their desks, doing most of their work for them, while they indulge in virtual reality games.
- Fixing nerve connections. In many cases it is outrageously stupid that people should be unable to move their limbs or to see, or hear, just because there is a little gap in the relevant nerve connection. We can fix electric cables, so fixing nerves should be routine by 2023.
- Replacement of simple organs. Never mind whether it's going to be artificial or biological, nobody will be able to tell the difference any more (there is my old "biotronics" sermon again!). For organs such as the kidney or the heart, whose function can be easily tak-

en over by a machine outside the body, there is no good reason why it should not be taken over by a new machine (engineered or grown) inside the body. The liver may take a bit longer, and expanding the approach to the brain would create an identity crisis, so let's leave that to the next generation.

- Exciting new materials. Whether it's spider silk or carbon nanotubes, I want new materials that open up new possibilities, such as space elevators or artificial organs.
- Understanding the Universe. Purely for curiosity reasons, I'd like to know what makes up those 95 % of the Universe which we currently don't have a clue about. As a fringe benefit, we might also find out about the future fate of the place. Not that it matters, as we won't be around long enough to witness any significant changes of the large-scale structure of the Universe one way or the other. But I do find it embarrassing for us as a civilization, having to admit that we don't understand the place we live in.

Of course it's quite possible that none or only a few of these things will actually happen. But many other things will, and there will be plenty more crazily unexpected discoveries, irresistibly sexy insights, and blindingly cool innovations to be reported in the coming years, and that's what keeps me going.

Index

The Birds, the Bees and the Platypuses. Michael Gross
Copyright © 2008 WILEY-VCH Verlag GmbH & Co. KGaA, Weinheim
ISBN 978-3-527-32287-9